AF443079

The Newcomen Engine in the West of England

THE NEWCOMEN ENGINE
IN THE WEST OF ENGLAND

by K.H. Rogers

MOONRAKER PRESS

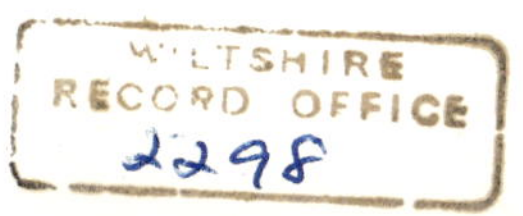

First published 1976 by Moonraker Press
26 St Margarets Street, Bradford-on-Avon, Wiltshire
SBN 239.00157.5

IBM Composer set by Roy Day Associates
Printed by Redwood Burn Ltd, Trowbridge
and bound in Great Britain

PREFACE

This book began as a short essay aimed at revealing the account book of Thomas Goldney as a source for the early history of the steam engine. This remains its basic purpose, but it has now expanded to include other material relating to Newcomen engines in Gloucestershire, Somerset and Cornwall. For the first two counties this comes from archive-sources hitherto largely unused in this context. Distance made a search in Cornish archives impossible, but far more work has been published on the history of mining and engines in that county, particularly by Mr Barton. I should be glad to hear of new references to engines, especially those mentioned as unplaced.

Mr H.L.Douch of the Museum at Truro helped greatly with the identification of Cornish mines. Thanks are due also to the County Archivists of Gloucestershire, Somerset and Wiltshire, and the City Archivist of Bristol, for permission to quote from and reproduce documents in their custody, and to their staffs for help in the location of material. Mr Roy Day of Keynsham kindly took most of the photographs.

109

January 13th 1749

Rec.d from the Dale of Richard Beard's Trowe

73 — 4 Battery Carriages & 16 Wheels for Warmly
Compa.y. 16..–.25 @ 4/ ℔ Ⓕ ————— £ — 8.. 18.. 5½
1 Table Frame for m.r Perkins on y.e Buck @ —— 2.. 12.. 6
Sent m.r Lemon — Rec. another but return'd it, being broke } — 3.. 00.. 0
again mended say @ £3 Painted

Jan: 23.. — Mehetable Woods D.r to Thom.s Ingram for my
72 — Dra.ft on him to her for £28.. 5.. 9 in full of her } £28 — 5.. 9.
acco.tts I sent her 9ber 25..th 1749

Jan.y 20.. — Willy Compa.y D.r to Thom.s Ingram for
72 — Richard Ford's Dra.ftt on me at his house to } £100.. —..
Abra.m: Ford for £100 @ 2 d.os date

Jan.y 22 — Rec: from the Dale for W.m Lemon Esq.r of Beard's Trowe
y.e Invoyce date forwarded hitherto y.e care of Jos: Loscombe.
1 Cylinder 54 Inches Diam. 67..–21
1 Bottom for it ——— 18..2..18 – 85..3..11 @ 30/ } £128.. 15.. 5¼
1 Piston ——— 11..–10
1 Ditto with Stems — 11..–.21..22..1..3 @ 12/ — 13.. 7.. 3¾
2 wro.tt Iron Stems to y.e latter 2..3..5 at 5½ ℔ tt — 7.. 3.. 5½
73 — For welding them in at 12/ each — 1.. 4.. —
1 Brass Piston Ring — 2.. –.105
1 Ditto Sinking Clack — 1.. –.21 3..1..3 @ 16:/ ℔ tt 24.. 9.. 4
For Polgooth, to Padstow — £174.. 19.. 6½
1 Pipe 8 f.t long, 17 In Diam: 8..3–22
2 Ditto, 8–3..18, & 8–2..6, 17..1..24
For North Downs, to Hayle — 26..1..18 at 18/ ℔ Ⓕ £23.. 15.. 4½
£198.. 14.. 11

Feb.y 5.. — Willy Compa.y D.r to Thom.s Ingram for
72 — Richard Ford's Dra.ft on me at his house
to Abraham Ford @ 2 d.os date for — £200.. —

Feb: 9th — Rec: from Willy of Rich.d Beard's Trowe
74 — A fell of Pigg Iron sent for 10 Ton, but a eigh
as follows, & sold to Jas: Hillhouse & Compa.y
Feb.y 20.. — Gross 201.. — 11 break: 2..2..21, n. 198..1..18 @ 7/6 ℔ Ⓕ £74.. 8.. 1.

Sundry acco.tts are D.rs to Thom.s Ingram for Comm.n
& Postage he charges in his acco.tt sent me to Feb.y 15..1749
72 — viz £13..5.. for Commiss.n & 33/ Postage, in all £14..18..
Whereof £7..16.. for Dale Co.s Acco.t, £3..9.. — Willy Co.s
£1.. 5.. Leighton Co.s, & £2..8.. for my own acco.t — £14.. 18.. —

Typical page from Thomas Goldney's account book

THOMAS GOLDNEY

Thomas Goldney (1694–1768) was a Bristol merchant and banker, son and grandson of previous Thomas Goldneys who had been prominent citizens since the middle of the seventeenth century. The family belonged to the important Quaker community in Bristol, and it was probably this that led the second Thomas to invest money in the ironworks started by another Quaker, Abraham Darby, c. 1707. Darby soon moved from Bristol to Coalbrookdale, but the Goldney connection with his business remained strong. Thomas II owned ten of the 16 shares into which the company was divided until his death in 1731, and after the redistribution of 1738 Thomas III owned three-sixteenths and his six brothers and sisters one each. Father and son both acted as agent for the company in Bristol.

Thomas Goldney III never married. He devoted much attention to his fine house and grounds at Clifton, which became a well-known feature of the city. After his death his property passed in succession to a sister, then a brother, and finally to another sister who died in 1796. The Coalbrookdale shares were sold in 1773, and in 1796, because none of the third-generation Goldneys left issue, the Bristol property went to another branch of the Goldney family at Chippenham in Wiltshire, the birthplace of Thomas Goldney I.[1]

It is for this reason that an account book kept by Thomas Goldney III is now among the archives presented by the late Sir Henry Goldney to Wiltshire Record Office at Trowbridge. Its reference number there is 473/295, and details from it are reproduced here by kind permission of the County Archivist.

The book runs from 1 February 1741/2 to 19 January 1769, and except for the last few entries made after his death on 28 December 1768, is entirely in the hand of Thomas Goldney. Although the greatest number of entries are concerned with Goldney's agency for Coalbrookdale, it is essentially a record kept for his own use rather than a Company record, for it contains accounts of personal dealings with relatives over legacies and debts, and of his own shareholding in the Company and in other concerns. These included the Warmley Brass Co., Gronant lead mine (Flintshire), Kelyn mine, a mine at North Molton in Devon, Wheal Trugo in Cornwall, Shally Coghlan lead mine in Ireland, and various ships. Until 1748 there are more entries for guns and shot

This Agreement made the ...
of our Lord one thousand seven hundred and eighty ...
Henry Lord Middleton Thomas Smith Esquire and Mr ...
Westerleigh in the County of Gloucester on the one Part ...
of the City of Bristol and Mr Thomas Palmer of the Par...
of Gloucester (Engineer) on the other Part **Witn...**
Bond and Mr Thomas Palmer have jointly **Agreed** an...
to and with the Lords of the Manor of Westerleigh aforesai...
finish according to the Model and other Directions to tho...
Engine now erecting upon an Estate called Serridge bein...
Manor belonging to the said Lords And from the Day of th...
without Lett or hindrance to execute every necessary Bu...
Directors and Managers untill the whole of the said Engin...
proper Methodical and workmanlike manner so as to make it a good Engine, ...
every part of the said Engine on or before the first of October one thousand ...
during the Time herein limited will do their utmost to put and keep in ...
Coalpitheath belonging to the said Lords so that they may be used wit...
William Bond or his Partner Thomas Palmer shall constantly attend th...
and keeping in repair those others now used as aforesa...
and the other shall have his option to attend at such time and times as may...
expedient **In Consideration** of receiving from the said Lords the...
for each one according to his attendance and is in proportion for a...
also for the further Consideration of receiving upon compleating the whole ...
by them jointly received **But** in Case any Impediment shall arise which...
so as to hinder the possibility of carrying this design into Execution Then ...
to be freed from this Obligation and the said Mr William Bond and his ...
receive Ten Pounds of the said Lords for the Injury sustained by th...
hereunto set our Hands the Day and Year first above written

Witness to the Signature of
William Bond & Thomas Palmer

Stephen George

ifth Day of June in the Year
Between the Right Hon.ble
Colston Lords of the Manor of
nd Mr William Bond
of Stapleton in the said County
seth that the said Mr William
DO hereby Covenant and Agree
O Erect carry on and compleat the
work by Mr Robert Barber an
rt of the Freehold of the said
date hereof to begin upon and
s coming within their province as
finished and set to work in a
ther hereby Agreeing to finish
hundred and ninety And also
in the two Engines now used at
sets and that either of the said
ting and compleating the said Eng.
during the Period above mention
hem be judged necessary and
n of Twenty Shillings per Week
longer or shorter time And
another Sum of Twenty Pounds to be
Parties hereto cannot prevent
d in that Case each Person
Mr Thomas Palmer jointly to
In Witness whereof we ha
Bond
Palmer

g an engine at Coalpit Heath, 1789

than for any other product. Some appear to have been sent down to meet specific orders; thus in 1744 Captain Richard Prankerd, owner of the *Rover* privateer, took delivery of eight short nine-pounders which were hauled away from the head of the quay in Bristol, and for which he paid £213. In 1745 another privateer called the *Bristol* was supplied with 12 twelve-pounders, 23 nine-pounders, 600 twelve-pound shot and 1200 nine-pound shot, which cost in all £947. Other consignments were sold to firms which took them away to London or America for re-sale. In 1744 twenty half-pounders were sent to Bath 'for the use of that city on rejoycing days'.

A wide variety of domestic ironware also passed through Goldney's hands including pots, kettles, pestles and mortars, stoves, furnaces, fire-backs and flower-pots. There were also occasional ornamental items, such as the parcel of pallisading delivered to Vickeris Dickinson in 1747, which consisted of 149 bannisters, 158 spear-heads, seven pillars, seven pillar-heads, and eight rails. In 1753 the Society of Merchants in Bristol bought five turned pillar-cases for the Great Crane in the city. Industrial customers included the metal-working concerns around Bristol (for furnace parts and moulds), collieries (for waggon-wheels) and gunpowder mills (for iron beds); while in 1751 a notable consignment was of 107 water pipes 9 ft long and 5 in. in diameter sent to London for use in Kensington Gardens. Large quantities of pig-iron were regularly sent, some being sold to local firms and some shipped away again. Occasionally traffic was in the other direction, as in 1748 when consignments of 'pot brass' were sent to Coalbrookdale; one dealer supplied 162 pots weighing in all 26 cwt, and a hogshead of small brass, for which he was paid in all, at 7¼d. a lb., £141. The book also has considerable interest for the historian of inland nagivation, for it generally gives the name of the trow which brought goods down the Severn and of its master. Eustace Beard, the trowman whose iron gravestone dated 1761 is in Benthall churchyard in Shropshire, was a regular carrier of consignments to Goldney.[2]

An important series of entries refers to the delivery of castings for steam engines. The account book has, owing to its unexpected location, escaped the attention of historians of the Coalbrookdale Company and of the steam engine, and this book is mainly concerned with the infor-mation which it gives on the distribution of steam engines during the period it covers.

The early history of the steam engine has been the subject of much painstaking research. Mr Allen has made a list of all engines known to have been erected before the expiry of Savery's patent in 1733, which contains so far 82 certain, and a further ten possible engines in Great Britain, plus a few more abroad. The references come from a variety of sources, including the business records of the concerns which used them, and papers concerning legal disputes over engines in the Public Record Office.[3]

Probably the most important single source, however, is the records of the Coalbrookdale Company, which supplied engine-parts. In the eighteenth century a steam engine was by no means a unit supplied by one maker to be erected on the site; castings were brought from a foundry, but blacksmith's, carpenter's and plumber's work was obtained locally. Some simple cast-iron work might also be available from a local man, but the large castings, especially those which needed boring or turning, had to be produced by a specialist foundry. The very earliest engines had cylinders of brass, and as late as 1744 Desaguliers in his *Experimental Philosophy* recommended them on sound theoretical grounds; they could be cast with thinner walls than those of iron, and so heated and cooled, as the engine cycle demanded, more rapidly, which increased the rate of working.

Whether or not this was appreciated by the engineers, brass cylinders continued to be made until the 1730s, apparently at foundries near London. But the cost of a brass cylinder was great. Farey quotes a bill for an engine erected in County Durham in 1727 in which the 29-in. cylinder cost £250.[4] An iron cylinder of a slightly larger diameter cost £49-15s. in 1745 (see below p.41), so it is not surprising that the iron cylinder was quickly introduced, and became more and more common as engines grew larger.

Iron cylinders were known as early as 1718,[5] that is within six years of the erection of the first successful engine at the Coneygree Coal Work, Tipton (the Dudley Castle engine of contemporary writers). Four years later the first known iron cylinder was made at Coalbrookdale and from that time the company's surviving records have been the most important single source for the distribution and size of engines in use between 1722 and 1748. Dr Raistrick has described the nature of the records fully in Chapter 8 of *Dynasty of Iron Founders.* Some of the information comes from payments recorded in cash books to workmen for actually casting and dressing the cylinders and other parts, while the stock books record the delivery of some parts to customers. Dr Mott has used these two types of record to make a list of 38 cylinders supplied between 1722 and 1738, and of a further 29 possible cylinders 1738-1748, when the form of the accounts makes it more difficult to discover what was being done.[6]

After 1748 the company's records refer only to engines erected for their own use, and, although there are other sources such as the few surviving business records of the users and newspaper accounts and advertisements, the record is naturally less complete. Early references to the use of engines in Cornwall are particularly short and equivocal. Allen's list includes only four certain Cornish references, and three more possibles, before 1733, while Mott found no clear Cornish references up to 1748.

Between 1744 and 1768 Goldney recorded details of 49 cylinders which passed through his hands, giving in all cases the name of the buyer and usually the dimensions and weight and the place where the engine was to be erected. But the detail given in his entries is not uniform, and a number of other places where engines were in use can be deduced from the delivery of castings other than cylinders, or of 'parcels of fire engine castings'. This brings the total number of engines dealt with by the account book to over 60. Some of these were known from other sources, but a number are hitherto unrecorded. The importance of this account book as a new source for the Coalbrookdale Company's activities in making engine-parts is best seen by comparing it with Dr Mott's list for the period when it overlaps the company's records, 1744 to 1748. None of the 14 cylinders which passed through Goldney's hands in those years appears in the Mott list.

The agency of Thomas the younger may have been on a formal basis and geographically defined. It is noteworthy that the agreement for the common pricing of cast-iron engine-parts which John Wilkinson, Abraham Darby II and Isaac Wilkinson came to in 1762 excluded South Wales, Cornwall, Somerset and the Bristol and London regions.[7] These were exactly the areas which Goldney dealt with, the only exceptions being one small engine sent to Scotland and one consignment of unusual size to Newcastle-upon-Tyne. The prices in the agreement were exactly the same as those recorded by Goldney throughout the book, so the significance of the exclusion is not clear. The largest number of customers was from the copper mines of Cornwall, followed by the collieries and other industrial concerns near Bristol. For these areas some further details of other known Newcomen engines are added. The other references in the book to parts supplied to South Wales, Newcastle-upon-Tyne and Scotland have also been included. The topographical survey is preceded by a short account of the Newcomen engine and followed by some more detailed analysis of the various parts supplied. An appendix gives some extracts from the account book *verbatim.*

Many descriptions of the Newcomen engine (or to use its eighteenth-century name, the fire engine) are available. A good one is in Farey's *Practical Treatise on the Steam Engine,* first printed in 1827 and now available in a modern reprint. Only the simplest description is attempted here, to make the following pages intelligible to the general reader.

The engine consists of a vertical cylinder, open at the top, but bolted to a bottom shaped like part of a cone. A piston moves up and down in the cylinder, and from a rod attached to the piston's upper face a chain runs up to one end of a large beam of timber pivoted at the centre. The ends of the beam are in the shape of the segments of a circle, called the arch-heads. From the arch head at the opposite end to the piston one or more chains run to the pump-work; the engines dealt with in this book were all pumps, for adaptation of the fire engine to provide rotary motion came later and was never common.

The cycle of the engine's working is as follows. Steam at low pressure is introduced into the cylinder from the bottom end through the steam valve. It drives any air out of the cylinder through a snifting valve at the side of the cylinder, and its pressure is sufficient to lift the piston to the top of the cylinder, the weight of the pumpwork at the opposite end of the beam assisting in this. The passage of steam or air through the gap between the piston and cylinder was, before the days of accurate machining, prevented by a gasket of rope packed with tallow on the upper side of the piston, and it was also the practice to keep a few inches of water on top of the piston. If any leakage occurred it only led to water which would sink to the bottom, entering the cylinder, rather than air which would lessen the vacuum. When the piston is at the top of its stroke (and so the pump work at its lowest point), the steam supply is shut off and a jet of cold water is shot into the cylinder bottom with sufficient force to hit the piston and so be sprayed throughout the cylinder. The supply of cold water comes from a cistern near the beam supplied by a small pump called the jackhead pump, also worked by the engine. The jet of cold water condenses all the steam, so producing hot water which falls to the cylinder bottom, leaves the cylinder through a valve, and conducted by an eduction pipe (called here a sinking pipe) to a hot well for the boiler feed-water. The cylinder is now a vacuum and so the piston falls to the bottom of the cylinder lifting the pump work at the opposite end of the beam by the amount of the engine's stroke. The cycle then begins again. The steam and injection valves are worked by rods hanging from a smaller arch-head part way along the beam.

The pump-work consists of a series of flanged pipes or pit barrels running down the shaft either from the top or from adit level, an adit being a tunnel by which water flows away to the nearest valley bottom. The pump-rod (or spear) runs in the pipes down to a working barrel, which is a machined pipe placed some feet above the well of the mine. The end of the spear works up and down in the working barrel, and to it is fixed a bucket which consists of an iron ring divided into two by a diametrical bar. To this bar are fixed semicircular flaps of leather, which are lifted up into a vertical position by the water when the spear is falling. As soon as it begins to rise, however, the flaps fall back to the horizontal position, so that the bucket then lifts the column of water up through the pipes. Below the working barrel is another pipe called the clack pipe. The clack is a flap valve similar to that on the bucket, but fixed in the pipe. When the spear is falling the clack is shut and holds the column of water in the pipes stationary. When the bucket is rising however, suction draws water into the lowest pipe of all, which is actually in the well of the mine; this is called by Farey the windbore, a term which appears in the account book sometimes, but its more usual name was blast-hole pipe. The same suction opens the flaps of the clack and allows water to rise into the working barrel. To avoid the necessity of lifting the whole of the pumping rods out to attend to the bucket, the pit barrel next above the working barrel has a bucket door, while the clack pipe has a clack door (sometimes at the lower end of a short separate pipe) to enable the clack leather to be renewed.

Bristol July 1793 —

Mess.rs Haynes & Co.

Bo.t of Harfords & Bristol Br.s. Co.

materials of a Fire Engine as P.r Valuation

Weigh Beam w.th Iron martingals Gudgeon &c		35.0.0
Cast iron Chains with w.t Iron Bolts	17.1.26 24/	5.4.10
Injection Cistern Lead on the inside Cylinder Caps & all the Lead Pipes	43.2.22 19/	39.6.10
Jack Pump Beam & w.th Iron work		0.10.6
Roof over the Engine House Timber & Tile		3.0.0
2 Inside Spring Beams		2.0.0
1 Plug		0.5.0
all the w.t Iron belonging to the Engine & Boylers	38.2.26 13/	25.3.6
1 Piston & Stem together	15.0.0 9.6	7.2.6
1 Cylinder	68.1.10 14/	47.16.9
1 Cylinder Bottom	18.3.18 9/	8.10.3
1 Fish Well	5.2.0 9/	2.0.6
1 Steam Receiver	12.0.13 4/	2.8.6
all the Steam Pipes, sinking Pipes & feed Pipes	56.3.23 8/	22.15.8
Brass Cocks & Regulators belonging to the Engine & some Brasses in the Counting House	5.3.25 4/	16.14.6
2 damper Plates	5.0.21 7/	1.16.1
1 Spring Carriage		3.0.0
1 working Barrell for the Jack Pump	10.0.0 14/	7.0.0
1 Clack Piece & Brass Clack together	6.3.0 9/	2.14.0
A old wooden Pipes with w.t Iron Hoops		1.0.0
1 Capstan w.t Iron, Brass Timber & Tile over the same		8.0.0
2 Cylinder Beams		2.0.0
2 Y Posts & Blocks		0.5.0
1 Engine House wooden Door & frame		0.5.0
A Engine Window frames with some Glass		1.10.0
2 Fire Engine Door frames & 1 Door	3.1.0 7/	1.2.9
2 w.t Iron Boilers with Copper Pipes Valved		70.0.0
1 Damper Frame & Wheel		0.5.0
2 Roofs over the Boilers Timber & Tile		5.0.0
Sundry pieces of old Timber		1.0.0
1 Screw Jack		1.10.0
carried over		£324.16.5

Bill for an engine moved to Cowhorn Hill, Bitton, 1793

In early engines the boiler was placed immediately under the cylinder and was generally constructed of copper with a lead top. The account book shows that boilers of iron plates were common in the 1760s, and also refers to a brass boiler for a fire engine made for the Ketley Company by the Warmley Brass Company in 1760 at a cost of £402. Goldney himself was supplied with two cast-iron boilers for his own small engine.

CORNWALL

In 1716 the proprietors of the Savery patent issued an advertisement which said that engines had been built in the counties of Stafford, Warwick, Cornwall and Flint.[8] The Cornish engine (or engines) has never been certainly identified. Traditions refer to early engines at 'Balcoath' near Porkellis in Wendron, where steam is said to have been raised by burning turf, at Wheal Vor in Breage[9] and at Wheal Rose near Truro (ie. in North Downs). The first engine in the county accepted by Mr Allen is attributed to Wheal Fortune in Ludgvan in 1720, but in his addendum he adds details which certainly belong not to that early engine but to one supplied to a mine in Ludgvan in 1746 (see below p.23)[10] Other firm references come from the mid-1720s, to engines at Wheal Rose, Wheal Busy, and Polgooth.

The excessive fuel consumption of the atmospheric engine was always a great drawback in Cornwall; the engine at Wheal Rose is said to have been so expensive to run that the adventurers drove a mile-and-a-half adit so that they could discontinue it.[11] In 1741 only three were at work in the county; these have been presumed to be the ones at Wheal Rose, Wheal Busy and Polgooth. In that year, however, a concession was made to the mine adventurers by the remission of duty on coal brought by sea to work fire engines.[12] Dr Rowe, however, considered this of less importance than the increase of local skill in running the engines to best advantage. There had been no skilled engineers in Cornwall until the brothers Jonathan and Josiah Hornblower came from the midlands in 1745. When they began to set up engines the men who tended them were able to make minor changes which improved their running and their fuel consumption. According to Dr Rowe, the expansion of Cornish mining by driving to deeper levels using engines came about a decade later than the remission on duty, and after the peace of 1748 made conditions for shipping coal in and ores out free from the menace of privateers.[13]

The Goldney account book corrects this view, by showing that 11 cylinders were sent to Cornwall during the years 1744–50, compared with only five from 1751–60. The expansion must, by these statistics alone, have occurred quite soon after the drawback on fuel. The first list of Cornish engines is the rather unsatisfactory one in William Borlase's *Natural History of Cornwall,* published in 1758. It mentions 14 engines 'and some others', while the account book has details of 16 cylinders supplied up to and including 1758, plus clear references to an engine at North Downs probably erected before 1744. Comparison of the two sources in a table shows that they are not easy to reconcile completely:

Borlase	*Account book*
Ludgvan Leaze 47-in.	Ludgvan Leaze 47-in.
Drannack 70-in.	Drannack 70-in.
North Downs 1	North Downs pre-1744
North Downs 2	North Downs 60-in. 1756
Pittlouarn 1	
Pittlouarn 2	
Polgooth	Polgooth 1747
	Polgooth 54-in. 1750
Wheal Reeth in Godolphin Ball	
Herland	Drannack 55-in. 1748
Bullengarden	Dolcoath 40-in. 1746
Dolcoath	Dolcoath 54-in. 1753
The Pool	Pool Adit 1745
	Pool Adit 60-in. 1747
Bosproual	
Wheal Rose	Wheal Rose 60-in. 1753
	Roskear 47-in. 1746
	Sir Wm. Lemon 1746
	Sir Wm. Lemon 52-in. 1749
	Chacewater 54-in. 1750
	Wheal Virgin 60-in. 1758

Pittlouarn was later considered to be part of Chacewater or Wheal Busy, a Lemon mine, so no doubt the two engines there can be equated with two of those of 1746–50 attributed by the book to Lemon or Chacewater. Other discrepancies could, of course, arise owing to the removal of engines from mine to mine.

Another figure was quoted by Farey, that in 1770 there were about 18 large engines in Cornwall, of which eight had cylinders of from 60 to 70-in.[14] The sizes of the others were not stated, but were presumably smaller. The account book refers to 31 engines in the county, of which 12 had cylinders of 60-in. or over, so that Farey's figure appears valueless. If, however, we take the account-book engines and those mentioned in other printed sources, Pryce's statement in his *Mineralogia Cornubiensis,* published in 1778, that above three-score engines had been built, seems perfectly plausible. About that number are mentioned in this book.

At the time Pryce wrote, the Cornish copper mines had for ten years been suffering from very severe competition from the great deposits of copper ore discovered in 1768 at Parys Mountain in Anglesey. These were of lower grade than the Cornish ores, but so near the surface that they could be worked open-cast. The only way Cornish mines could survive was to cut costs, especially those of pumping, and it was at this time that the first generation of Cornish steam-engineers arose. John Budge in particular erected several Newcomen engines which won some measure of approval from James Watt, and the foremost national engineer of the day, John Smeaton, was also brought down to the county. The large engine he built at Wheal Busy in 1775 probably represented the zenith of the Newcomen design.[15]

Within three years, however, the advent of Boulton and Watt, bringing Watt's more economical engine, was providential for Cornwall. As Pryce wrote, they had already set a small engine to work at Wheal Busy, and were building three more large engines.[16] Within a year or two the fire engine was almost completely superseded; only one was left in 1783, and that was idle.[17]

The sections on individual mines that follow deal firstly with those in which the adventurers Sir William Lemon and his successor Thomas Daniel were concerned, then with those of Abel Angove, and finally with a miscellaneous group in chronological order.

WHEAL ROSE, NORTH DOWNS

There was an engine on Wheal Rose by 1725, but it was 'so very chargeable' that the adventurers made a 1½-mile adit so that they could do without it.[18] The mine was very rich, however, and an engine on it was one of only three at work in the county in 1741.

Sir William Lemon and Co. in 1753 paid £194 for a 60-in. cylinder and bottom (99 cwt together), a piston with a wrought-iron stem, a sinking-pipe in three parts and a maundrell. Twenty-one pipes and three plates, which cost £231, followed early in 1754.

WHEAL BUSY OR CHACEWATER

An engine had been erected on this mine by 1727[19], and it was one of the three mines which had one working at the time that the duty on fire-engine coal was remitted in 1741. The first parts which the account book refers specifically to Chacewater were two pipes and a bucket-door plate amounting to only £10 in 1748. The adventurers were William Lemon and Co.

In 1750 were consigned the parts of an engine for Chacewater. They included a 54-in. cylinder weighing 68 cwt., a bottom, two piston-plates, a brass piston-ring, two brass pipes, three bucket doors and 12 pipes. The total cost was £359. Subsequent consignments marked specifically for Chacewater were sent in 1753 and 1754, and consisted almost entirely of pit barrels, 32 in all, at a cost of £265.

No more is known until 1766, when the mine was worked by Thomas Daniell and Co. They were sent a cylinder 66-in. by 10 ft., weight 96 cwt., a bottom, two sinking-pipes each in two parts, two working barrels, a jackhead, 35 pit-barrels, 34 pipes of various kinds and a few small parts. The cost, with one parcel of £82 unspecified, was £1016. Later in the same year and early in 1767 three more unspecified parcels were sent for Chacewater, at a further cost of £552. It seems likely that they included parts of another engine, probably with a 64-in. cylinder, for two cylinders of 66 and 64 in. with pistons, bobs (i.e. beams) and other parts were offered for sale on the mine in 1770. [20] Farey said that the mine was worked with two engines before Smeaton built his 72-in. engine on it in 1775, but he gave the diameters of the cylinders as 64 and 62 in., and said they only worked a six-ft stroke, which seems unlikely. He also said that one of the engines pumped the bottom 24 fathoms of the mine, and the other the next 26 fathoms needed to bring the water to adit level. The pit-barrels supplied in 1766 amounted to 45 fathoms, which with the working-barrels and blast-hole pipes would make the full length. But it is curious that half the pit-barrels were 20 in. in diameter and half ten-in. With the arrangement Farey mentioned, it would seem that they should have been about the same size for both lifts, not one lift much larger than the other. Farey says that the diameters of the lifts were 18½ in. and 17½ in. respectively. He also says that the adit was 25 fathoms deep; possibly the small pit-barrels enabled one of the engines to raise the injection-water for both engines to the surface.

It was on this mine that one of the most celebrated of all Newcomen engines was erected, that designed by John Smeaton in 1775. Farey gives a very full description of it. The cylinder was 72 in. in diameter and weighed 96 cwt., and the bottom weighed 33 cwt. It worked three columns of 16¾-in. pumps with a stroke of 9 ft, the total height being 51 fathoms to adit level. At nine strokes a minute this worked out at 76 h.p., which made this engine the most powerful built at that time. [21] After working for a few years it was altered to Watt's improved system; the cylinder was retained as an outer casing for a 63-in. cylinder cast by Wilkinsons. [22]

POLGOOTH

Like Chacewater, this mine was worked by William Lemon and Co., and had had an early engine on it. In 1747 parts of an engine for Polgooth were sent down to Joseph Loscombe. The combined weight of the cylinder and bottom was 63 cwt., which suggests a diameter of about 40 in. Other parts included three brass barrels, 42 iron barrels, a sinking-pipe and two pistons. The total cost was £645.

Three years later a 54-in. cylinder and bottom, two pistons with stems, a brass piston-ring and a sinking-clack, cost the company another £174. Consignments to Polgooth continued until 1754; the largest of these consisted of pit barrels, of which twelve 6 ft. x 15 in. in size went in January 1751, and 18 more of unspecified size in November of the same year.

NORTH DOWNS

This mine was in full production by 1740,[23] and regular consignments were recorded by Goldney between 1747 and 1754. They mainly consisted of brass and iron barrels and pipes. In 1756 a new engine was sent. It consisted of a cylinder 60 in. by 10 ft (weight 93 cwt.) with bottom and turned piston with stem, a sinking-pipe, a snifting-pipe, four gudgeons and four side plates, amounting in all to £233. In 1759, 26 pit barrels cost the company £203 and minor pipes and plates a further £36.

In 1758 two engines on this mine were mentioned by Borlase, of which one no doubt dated from before 1747 and the other was that of 1756.

WHEAL VIRGIN AND WEST WHEAL VIRGIN, ST. DAY

In August 1757 an unspecified parcel amounting to £220 was sent to William Lemon and Co. for Wheal Virgin. Early in 1758 the company paid £229 for a cylinder 60 in. by 9 ft (89 cwt.), a bottom, a piston-plate with two stems, a sinking-pipe in two parts and a steam-pipe. Judging by this price the parcel of 1757 could also have been the principal parts of an engine. Borlase, whose book was published in 1758, referred to Wheal Virgin as having an engine.

Further consignments of pipes, pit-barrels, and other small parts, and some unspecified parcels, followed regularly from 1758 to 1764. The only one large enough to include the main parts of an engine amounted to £247, in February 1761. About that time the principal adventurer became Thomas Daniel, whose company certainly fitted another engine to Wheal Virgin early in 1765. The cylinder was 60-in. by 10 ft., and weighed 80 cwt., but no other parts were specified. The bills between December 1764 and February 1765 came to £483.

In 1767 another cylinder of exactly the same dimensions and weighing 82 cwt. was sent, with a bottom, a piston, two wrought stems and a brass piston-ring, three working-barrels and 29 pit-barrels. This with another unspecified parcel came to £706. Further small parts continued to be supplied until the end of the account book in 1768.

In 1767 two small parcels of engine castings were sent to Daniell and Co. for West Wheal Virgin. It was on Wheal Virgin that the last Newcomen engine in Cornwall, by John Budge, the Camborne engineer, was built in 1778. In 1782 Consolidated Mines was formed by the amalgamation of Wheal Virgin, West Wheal Virgin, Wheal Maid, and Carharrack Mines, and the seven fire engines on them were replaced by five from Boulton and Watt.[24]

WILLIAM LEMON AND CO., UNSPECIFIED MINES

We have already mentioned engines supplied to five celebrated mines in which Lemon was the principal adventurer. Between 1745 and 1759 the account book also records much material sent down for him but with no mine mentioned Besides much pitwork and small parts it included two consignments which contained the main parts of engines. The first was in March 1746; the cylinder weighed 48 cwt., so was probably about 40 in. in diameter, and with its bottom, two pistons and sinking-pipe came to £110. Three years later a 52-in. cylinder (83 cwt.) with bottom, piston and brass piston-ring, a sinking-pipe in two parts, a sinking-clack, an injection cock, a snifting clack and an injection cap, amounted in all to £202.

Another unspecified group of castings, sent in August 1756 and costing £322, could possibly have represented an engine.

Many of the entries relating to Lemon's consignments mention a certain Joseph Loscombe; that he was perhaps simply an agent rather than an engineer is suggested by an entry of 1748 which shows that a consignment to Loscombe was for Joseph Hornblower.

Thomas Daniell began as Lemon's agent, succeeded to his business interests, and prospered on his own account, so that he became popularly known as 'guinea-a-minute' Daniell. We now mention a group of mines in which he was principally concerned.

WHEAL SPARNON

Consignments were made for this mine between 1760 and 1767. Those specified included much pitwork, but a parcel of fire-engine castings in July 1764 came to £413, which almost certainly represents the erection of an engine. Mr Jenkin tells us that there were two engines on the mine in 1765, one erected by Jonathan Hornblower.[25]

POLDORY, GWENNAP

The first recorded consignment for this mine was in July 1761, and included a piston with stems. It seems likely that an unspecified parcel of fire-engine castings at £245, which followed in September included a cylinder, although as there was an engine on the site by 1758, this is not certain. Another large unspecified parcel of fire-engine castings, amounting to £360, was sent in 1764, and this also seems large enough to represent the erection of an engine.

Regular consignments of small parts and pitwork were made until 1768, and Captain George Simmons received a 73-gallon furnace as a present.

ALE AND CAKES, ST, DAY

In March 1763 47 pit-barrels were sent to this mine, followed in April by unspecified fire-engine castings amounting to £296, so again the erection of an engine at this time can be suspected. The only other large parcel sent was one of £243 in 1767.

HEWAS MINE, ST MEWAN

Two engines on this mine were offered for sale in 1774.[26] Neither of the two recorded consignments, in 1763 and 1764, was very large. The only specified parts were a working barrel, a jackhead, three pipes, and three door-plates.

POLDICE, GWENNAP

Daniell and Co. were supplied with parts for an engine on this mine in 1763. The cylinder was 63 in. by 9 ft 9 in. and weighed 81 cwt, and with it were supplied its bottom and piston, and a large quantity of pitwork and pipes. The total amount sent to the mine in that year cost £850. Several small quantities were sent later, until 1768.

Mr Barton quotes an advertisement of 1765 for shares in Killycor and Poldice 'on which one fire engine is lately erected and another one building'. The former was presumably the 63-in. of 1763.[27] James Watt remarked of the two Poldice engines that they were 'among the best in the county of that species', and one was apparently still at work in 1780.[28]

In 1767 two small consignments, one including a working-barrel, a jackhead barrel, and seven pit-barrels were marked for Little Poldice.

TRESAVEAN MINE

Borlase reported one engine on this mine in 1758. The only entries recorded by Goldney all come from 1765, and include a cylinder, 60 in. by 10 ft, weighing 84 cwt., its bottom, piston, and brass piston-ring, quantities of pipes and barrels, fire-door frames and bars, a jackhead working-barrel and other parts. The cost in all was £829.

'WEST GAMBLER COPPER MINE', PROBABLY GRAMBLER IN ST AGNES

The only item for this mine was a spigot-ended working barrel 8ft by 10 in. in 1766.

'PENNALS MINE', PENHALLS, ST AGNES

The only specified item for this mine was also a spigot-ended working barrel, 9ft by 11 in., sent in 1768, but a small unspecified parcel worth £53 went in 1766.

'PELBARRA MINE', POLBERRO, ST AGNES

Three cranks were sent to Daniell and Co. for this mine in November 1768, at a cost of £18.

THOMAS DANIELL AND CO., UNSPECIFIED MINES

Three unrecorded parcels were sent between January and July 1761. The two largest, of £63 and £314, were noted as fire-engine castings, so they may represent the erection of an engine.

TREVENSON OR POOL ADIT, ILLOGAN PARISH

On the last day of 1744 Goldney delivered to Thomas Lanyon for Abel Angove and Co. of 'Tevenson in the parish of Logan' a bored cylinder weighing 43 cwt., a bottom, and five barrels, costing in all £120. The weight of the cylinder suggests a diameter of about 40 in., but on 2 February 1745 it was sent back to Coalbrookdale having been found to be faulty. On the same day a further consignment was sent to Cornwall, this time to Sir Francis Vivian and Co., and subsequent entries make it clear that this was the same concern. This and a later consignment consisted of pitwork, and then in October another cylinder was sent, with more pitwork. The cylinder weighed 60 cwt. and was presumably a replacement for the faulty one, but no bottom or piston was mentioned. More pit-barrels and pipes followed in 1746, bringing the expenditure to £801.

Two years later, a larger engine was fitted on this mine, now called Pool Adit. The parts which were delivered to Thomas Lanyon consisted only of a 60-in. cylinder weighing 90 cwt., and its bottom, and cost £198. A large amount of pitwork which followed included pit-barrels, two brass barrels, and other pipes and doors. The cost of this was £875.

The only further item specifically described for Pool Adit was a single brass pipe in 1751.

HERLAND OR DRANNACK, GWINEAR

In 1748 Abel Angove and Co. were the recipients (via Thomas Lanyon in Bristol) of a 55-in. cylinder weighing 71 cwt. and a bottom for Drannack. Again, as on other Angove mines, no piston was mentioned, and the cost of the two parts was £154. A sinking-pipe, three brass working-barrels, and 41 pit-barrels came to £651. The only other recorded material sent to Drannack was a small consignment in 1751.

Borlase, however, refers to a 70-in. engine on Herland, the more usual name for this mine, and this can be identified with a cylinder 70 in. by 9 ft 8 in. consigned to Angove, with no mine mentioned, in May 1753. It weighed 116 cwt. and with its bottom cost £247. The large quantity of pitwork sent to Angove in the same year could well have been for Herland. It included 31 pit-barrels, a working-barrel, and eight pipes, and cost £574. A piston sent in October 1753 was also probably for the same engine.

Borlase recorded two engines on Herland in 1758, and two were offered for sale in 1767, one the 70-in. and the other of 60-in., so that the 1748 engine had apparently been replaced at some time. The same advertisement mentions a 63-in. engine on Drannack Mine,[29] but this cannot be identified in the account book.

Between 1754 and 1765 regular small parcels were sent to Abel Angove and Co., but no specific mine was mentioned.

LUDGVAN LEAZE, LATER MOVED TO GREAT WORK

In 1746 Joseph Percival and Co. were supplied with a 47-in. cylinder and bottom, a sinking-pipe, an iron barrel and a few other small parts for Ludgvan Leaze at a cost of £134. No piston is mentioned. The company was allowed a rebate of £287 for supplying bell-metal for casting their working-barrels for this mine and for Roskear, but no details of shipping the barrels are entered.

In 1754 the cylinder of the Ludgvan Leaze engine was moved to Great Work in Godolphin Bal. The firm there was John Rogers and Co., who were supplied in that year with 16 pipes, 23 pit barrels and two working barrels, at a total cost of £304. The order came from the Cornish engineer John Nancarrow, who left details of the working of the 47-in. engine on the two mines. At Ludgvan Leaze it raised water from about 30 fathoms under adit, with pit barrels 15 in. in diameter, and made 14 or 15 strokes a minute, delivering a hogshead at each stroke. At Great Work the pit barrels were only ten-in. ones, but the engine drew from 50 or 60 fathoms below adit, and also had to raise the boiler feed-water from the 30-fathom adit level.[30]

Mr Jenkin quotes from a cost book of Great Work, 1759–64, which shows that George John, a Camborne engineer, cast various brass parts for the engine, but also records iron-work sent from Coalbrookdale.[31] The Goldney book mentions several consignments to Rogers and Co., mostly unspecified and none large, up to 1762. There was later a 63-in. engine at Great Work which was offered for sale in 1773.[32]

ROSKEAR, CAMBORNE

In 1746 Joseph Percival and Co. bought for Roskear castings amounting to £657. They included a 47-in. cylinder with a bottom and two pistons, a sinking-pipe, three brass working barrels, and 32 pit barrels. Seven more pit barrels and a middle length for a sinking-pipe were sent in 1758.

DOLCOATH

The earliest entries in the account book relating to Dolcoath, in July to September 1746, record the supply of a 40-in. cylinder weighing 46 cwt., a bottom and a piston, a sinking-pipe, two brass barrels, 15 pipes and two bucket-doors. The cost was £366. In 1748 a sinking-pipe and more pitwork, consisting of a brass barrel, ten iron barrels and a bucket-door plate, were supplied.

The adventurers named in these entries were John Freeman and Co. The next consignment, two pipes and a damper plate in 1750, was to John Jones and Co. who were engineers in Bristol. A further small shipment for Dolcoath in 1752 was sent to William Churchill and Co., and in February 1753 they were named as the consignees of parts of another engine. These were a cylinder 54 in. x 9 ft weighing 62 cwt., a bottom and piston, a sinking-pipe in three parts, four mouth plates, a damper-plate, a Y-weight and two small pipes, amounting in all, to £194. In July of the same year two working barrels were sent to John Jones and Co., and they continued to be named in several small entries until October 1755. The last of these was sent to the engineer John Wise.

Borlase's list only refers to one engine at Dolcoath, but the Bullen Garden engine of his list was at the shaft of that name on this mine. In the period from March to June 1768 a new series of entries refer to consignments to Francis Bassett and Co. of Tehidy for Dolcoath New Mine. A cylinder 63 in. x 10 ft weighing 83 cwt., its bottom, a piston, sinking-pipe, steam-pipe and three other pipes came to £257, and the company paid a further £560 for a great quantity of pitwork, including a working barrel and 63 pit barrels. An unspecified parcel of £41 was also sent.

Mr Harris published details from the Dolcoath pay books, which begin in April 1771, of payments to the engineers John Wise and John Budge, to the Coalbrookdale Co.(£203), and to the adventurers of Weeth Mine and North Downs, which he assumed related to the erection of the 63-in. engine. It seems more likely that they relate to repairs or additions to the pitwork, since they come three years after the supply of the engine parts.

Another engine was erected on the Bullen Garden shaft in 1775. Francis Trevithick attributed this to Richard Trevithick the elder, but Mr Harris has shown that the engineer concerned was John Budge of Camborne. The main parts were bought from the adventurers at Carloose Mine for £414, and other recorded payments included £131 to Coalbrookdale for iron pumps, and £139 to John Jones and Co. of Bristol for pumps and gudgeons. The total cost was about £2000.

James Watt inspected several of Budge's engines, which were reckoned to be the best in Cornwall. In 1777 he wrote that he had seen at Dolcoath and Wheal Chance five engines built by Budge of 60-, 63- and 70-in. cylinders. Three of these were at Dolcoath, and must have included the 1768 63-in., and the 1775 Carloose engines. Whether the two smaller and earlier engines were also still at work is not known.[33]

WEETH MINE, CAMBORNE

In 1754 Sampson Swaine erected a copper-smelting furnace at Entral in Camborne. In 1762 he obtained a patent in which two new ideas were included; one was to make a fire engine work a horizontal axle by means of two racks fixed to its beam, and the second was to use the heat of the smelting furnace for the boiler of the engine. In the first of these the motion of one of the racks presumably had to be reversed, perhaps by gearing, so that the axle always turned the same way. It is difficult to see how a regular motion could be obtained from an engine also used for pumping, and no more is known of this idea. The second part of the patent was, however, put into practice. In December 1762 Swaine and John Weston were granted a sett (mining lease) of the Weeth mine main lode. They were to erect 'their new invented engine with as many furnaces as may be wanted to work the same'. A covenant that they were to erect a fire engine to unwater the bottom, which was 20 fathoms below adit level, presumably refers to the same engine.

The account book mentions the consignment of 'sundry parcels' to Weston, Swaine and Co. in 1763, amounting to the large sum of £851, but the only parts specified were 170 boiler plates. The engine was erected by September 1764, when Swaine made an agreement for a water supply to it. Gabriel Jars, the French metallurgist, saw the arrangement c. 1768, and described the way in which the boiler was built over the furnace and also how tubes to carry heat were passed through the boiler. Jars criticised the arrangement on sound practical grounds, that the furnaces required irregular heating which did not give the steady supply of steam needed for the engine. Dr Rowe quotes a contemporary pamphlet which suggests that the principle was successfully applied on one occasion only, presumably here.[34]

No more is known of Swaine, but the large fire engine at the Weeth Mine was offered for sale in 1774.[35]

PEDNANDREA MINE, REDRUTH

Between October 1760 and April 1761 consignments amounting to £251 were sent to John Dudley for Pednandrea Mine. The only parts specified were two clack seats, two blast hole pipes, and five pit barrels.

OWEN VEAN, TREGURTHA DOWNS, MARAZION

The only reference to this mine comes from 1762, when a parcel of fire engine castings amounting to £251 was consigned to George Blewitt and Co. A Boulton and Watt engine was being erected on this mine when Pryce wrote in 1778. It was probably the earlier engine which gave its name to the still-existing Fire Engine public house in Marazion.[36]

WHEAL OULA IN NORTH DOWNS

Between July and November 1763 William Churchill and Co. of Redruth were sent a cylinder 70 in. x 10 ft, weighing 110 cwt. and other unnamed castings, amounting in all to £268. Other consignments during this period were a large unspecified parcel of £454, and pitwork including a working barrel and eleven pit barrels amounting to £163.

WHEAL FORTUNE

There were several mines of this name Mr Barton mentions advertisements of 1765, when Wheal Fortune, Ludgvan, was to be let and the takers were to erect an engine of not less than 60 in.; of 1770 for shares in Wheal Fortune, Redruth, where two fire engines had been erected (one recently); and of 1778 of Wheal Fortune at St Hilary with a 65 in. engine only four years old.

The account book contains no reference to a complete engine for a mine of this name, but only several small parcels between 1760 and 1766. Some were for Abel Angove and Co. and some for William Munday and Co. at Redruth. This, and the information above, suggests that they were for the earlier of the two engines at the Redruth mine.

These two engines were advertised for sale in 1771, both having 70-in. cylinders.[37]

HIGHER ROSEWARNE AND WHEAL GERRY, GWINEAR
In 1768 three parcels of fire-engine castings were sent to John Ennis and Thomas Harris of Camborne for these mines. One included three working barrels and seven pit barrels and amounted to £151, while neither of the other two, £66 and £59, which were unspecified, seems large enough to have included a cylinder.

Besides the advertisements mentioned above, Mr Barton has references to three other mines not met with in the account book:[38]

WHEAL REETH.   A fire engine to be erected in 1764.

WHEAL PARK, MARAZION.   A lately-erected 67-in. engine offered for sale in 1771.

TRELEIGH WOOD MINE.   Sale of materials of a 47½-in. engine in 1773.

There were also two engines at
WHEAL CHANCE, GWENNAP.   These were built by John Budge, and were taken as the standard by which the savings of fuel of Boulton and Watt engines were measured. Their consumption over a year, 1777–8, was found to be about two-and-a-half times as much as one Boulton and Watt engine doing the same work, 1780–81.[39]

GLOUCESTERSHIRE

It has been said that there was a fire engine working at Bitton in 1724.[40] This, however, has proved to be a misapprehension, as the following passage from a letter of Mary Dafter dated 12 February 1723/4 will show: 'he [Mr Seed or Esqr. Trye] have bene the chefe progotter [projector] of a nue ingen with Vallenten Flouer and two more, to be worked with oute horses drahing of it and they have now a complished it but it is not to be discouvered tell they have gote a pattorne [patent] for it which now thay be a ponit so wee shall here in a lettell time where it will ancor [answer]. Mr Flouar telled me it could be made for £50 ...'[41] The last sentence is conclusive that this was not a fire engine, which would certainly have cost several hundred pounds; in fact the patent which Valentine Flower obtained (No.469, 1724) says that it was for raising water without the help of fire.

Little work has been done on the detailed history of the various collieries, although record material for some is copious. The only 'statistic' available comes from Donn's printed map of the country round Bristol made in 1769, which shows ten engine-houses. These do not include three engines at Warmley and two at Soundwell known to have been working by then. Besides these five the account book refers to seven other engines whose sites are known fairly certainly, and to four others which were certainly in the Bristol area, although they could have been in Somerset. They are included in this section for convenience, together with seven other engines to which eighteenth-century references have been found. Some of these were probably among the four unplaced ones of the account book.

All the engines worked on collieries except those at the Warmley Brass works, which were used to pump back water for water wheels, and the small engine which supplied the fountains in Goldney's garden.

WARMLEY

The history of the Warmley Brass Company has been excellently written by Mrs Joan Day in her recent work *Bristol Brass*. It began in 1746, and in 1748 a 36-in. cylinder and some other parts of an engine arrived at Bristol. This engine was probably erected and found to be too small, for parts of a larger one began to arrive in April 1749, and in 1750 the old parts were returned to Goldney and sold to Nibley Colliery.

The purpose for which the engine was required was to pump water from the tail of a water-mill back up 18 ft into a pond so that it could be re-used again and again on the wheels. It evidently caused a stir when it was installed for it was described in the *Bristol Journal* of 30 September 1749

'We are certainly assur'd that the curious fire engine just erected at Bearshood-Hill near Birmingham (by Mr. Jos. Hornblower, engineer) for Mr. William Champion and Comp. at their brass-warehouse at Warmley is perform'd most accurately, and gives the said gentlemen the utmost satisfaction as well as to the curious — The machine is the most noblest of the kind in the world; it discharges upwards of 3000 hogsheads of water in an hour. The water is buoyed up by several tubes in a hemispherical form, and falls into the pool as a cascade, and affords a grand and beautiful scene'.

The account book shows that the parts were delivered between April and August 1749. The 48-in. cylinder weighed 67 cwt., and there was a bottom, piston, and a brass piston-ring, a sinking-pipe, a bored house-water tree, four working barrels, 18 pit barrels, and various buckets and clacks. The total recorded cost of the castings was £465. It was said that the total cost of the engine was £2000, and that it used £300-worth of coal a year.[42]

In 1761 the Warmley Company was enlarged when three local coal-owners, Charles Whittuck, Charles Bragge and Norborne Berkeley,became partners.[43] In the same year the parts of a large engine were delivered. The cylinder, 74½ in. x 10 ft weighed 128 cwt., and the bottom 51 cwt. The parts of the engine are set out in detail, and include four bored working barrels 9 ft x 29 in., four pit barrels 9½ ft x 29¾ in., a hot well, three engine-chains and three hooks, a piston, three wrought stems, a breeches pipe, a steam-vessel and top, a housewater pipe, a brass regulator, and various clacks and other small parts. The cost of all was £916. This was the engine which Sir Joseph Banks saw working when he visited the works in 1767. It made from nine to 11 strokes a minute, and raised 17 hogsheads at each stroke by means of four 30-in. pumps. This was over 9000 hogsheads an hour even at the slowest rate, more than three times the work performed by the 48-in. engine. Arthur Young described the Warmley works in his *Tour through the Southern Counties* published in 1768; he mentioned the prodigious fire engine, but exaggerated its capacity to an incredible 3000 hogsheads a minute. The 74½-in. engine evidently remained at Warmley until 1784, when its cylinder was among a quantity of engine-parts offered for sale to Matthew Boulton by the Warmley Company's successor, the Bristol Brass and Copper Co.

Sir Joseph Banks also saw another engine at Warmley, which he noted had a 60-in. cylinder.[44] This was possibly a mistake on his part, for the account book shows that an engine of which the parts were delivered between March and June 1765 had a 70-in. cylinder. The only working barrel noted was 9 ft x 57 in., so the engine probably worked only the one pump, although a much larger one than those in the 74½-in. engine. Total recorded cost was £673. In 1772 the cylinder, bottom, piston, and six chains of this engine were valued and sold to Sir Jarret Smith.

Apart from the three engines the account book records intermittent deliveries of small parts for them and of many parts for furnaces.

NIBLEY COLLIERY

Charles Bragge lived at Cleeve Hill in Mangotsfield, and was lord of the manors of Mangotsfield and Frampton Cotterell and half the manor of Bitton. He was one of the local coalowners who in 1761 joined the Warmley Brass Company.

In 1750 he made a lease of the right to work coal under his property in Iron Acton to Richard Stokes of Stanshawe Court in Yate, gentleman; Stephen Clarke of Chipping Sodbury, cheese factor; and George Webb also of Sodbury, maltster. These men also took coaling leases of the lands of three other owners in Iron Acton, and Bragge then joined them in a partnership agreement to work the whole property, and to erect an cost, which was to be paid back later.[45] This concern was known as the Nibley Colliery after the hamlet of that name in the parish of Westerleigh, although the works lay in Iron Acton.

The first delivery to the firm, in 1750, consisted of a 10¾-in. working barrel, a wind-bore blast-hole pipe, two buckets and two clacks, amounting to £24 in all. These were to be used with a 36-in. cylinder weighing 51 cwt., a bottom, and seven pipes which had been returned from the Warmley Company in favour of a larger engine, and cost £138. A pipe which was delivered in October 1751 was returned as being too late, so the engine was evidently at work by then.

In 1753 parts for a second engine were delivered to the company for Nibley Colliery. The 44½-in. cylinder was 9 ft 11 in. long and weighed 83 cwt., and with it was a bottom, a piston, a sinking-pipe, and a brass sinking-clack. The delivery cost £128. From then until 1766 the company took several deliveries of miscellaneous parts, such as pit barrels and pipes, clacks, steam-vessels and boiler-plates, worth in all £246.

Donn's map of 1769 shows an engine-house in Iron Acton, between the road from Nibley to Acton and the river Frome, which certainly belonged to this colliery. On Taylor's map of 1777 it is marked 'Old Engine'.

SOUNDWELL COLLIERY

A 26-in. cylinder weighing 24 cwt., a bottom, a piston, a sinking-pipe and two working barrels were delivered to Samuel Whittuck and Co. in 1750, at a total cost of £66. Six years later parts for another engine were sent for the same company, consisting of a cylinder 33 in. x 9 ft weighing 37 cwt., a bottom, a piston and a sinking-pipe, worth in all £83. Boiler-plates worth £19 were sent in 1764, pipes worth £78 in

1766, and an unspecified £86-consignment, various small parts worth £44, and 95 boiler-plates worth £40, all in 1767. Only the 1764 entry mentions Soundwell Colliery, so it is possible that the earlier engine was elsewhere.

Donn shows no engine at Soundwell, but Taylor's map of 1777 shows one engine-house there. The Whittuck family was an important one in the working of collieries, and in 1777 Charles Whittuck took a lease from Charles Bragge of coaling rights in the whole manor of Mangotsfield except the coal-works near Staple Hill.[46] At this time the family probably held, as it certainly did later, rights in the Newton family's liberty in Kingswood, which would have included Soundwell.[47]

HANHAM

In 1750 parts of an engine were sent for Moses Slade and Co. They consisted of a 28-in. cylinder weighing 27 cwt., with bottom, piston and sinking-pipe, two working barrels, two blast-hole pipes, two clack-seat pipes, a brass regulator, three clacks, two cocks, an injection-cap, and a jackhead pipe. The total cost was £118.

Other entries in the account book show that Moses Slade and Co. were in the lead-smelting business, buying pig-lead moulds and a lead pot, but the engine was probably for a colliery. No indication is given in the account book, but by a lucky chance we know fairly certainly where it was in 1793. In that year Haynes and Co. of Cowhorn Hill Colliery had valuations made of two engines, distinguished as 'big' and 'little',[48] and the big one, which they bought, was certainly at Hanham, (See p 39) The little one was almost certainly there too, and the supposition is supported by Donn's map of 1769, which shows two engines there. Certainly the little engine had the same cylinder and bottom as the engine supplied to Slade in 1750, although the possibility of it having been moved to Hanham from elsewhere cannot be ruled out.

COALPIT HEATH. WESTERLEIGH PARISH

In 1751 Jarret Smith and Co. were supplied with a 42-in. cylinder weighing 54 cwt. with bottom and piston, a sinking-pipe, two 7-in. bored pipes, two 8½-in. working barrels, 22 pit barrels, 12 buckets, four clacks and a brass clack, and two clack-doors. The cost was £307. No place was mentioned, but Smith's own papers provide the information. He was a Bristol attorney who by marriage gained Ashton Court and the great estates of the Smyth family, which included half the manor of Westerleigh. The Ashton Court collection at the Bristol Archives Office includes many bills and accounts concerning the coal works there.

The bill of a certain Robert Taylor runs from October 1750 to May 1751, and besides general haulage work includes several days' and one night's work at the engine in December 1750. This may refer to getting a place ready for the parts which began to pass through Goldney's hands in January 1751, although the night work suggests that there was an engine already at work. The Westerleigh manor coal account for 1751 gives the totals of seven bills 'about the fire engine'. The first was Goldney's £307, followed by Willetts and Wilding £60, James Malcott £37, Thomas Hill £88, William Tyly £3, Nathaniel Arthur £90, and James Hillhouse £82. The engine thus appears to have cost in all £668. Other bills show that Willetts and Wilding were of Framilode, near Stroud, and were supplying iron plates for the boiler. Hillhouse was in the iron-trade in Bristol.[49]

In 1763 Sir (as he now was) Jarret Smith was thinking of opening a new pit. 'We think' wrote Alexander Colston, who had a part share in the existing one, 'that your setting up a new coal work, if you proceed in it when we have plenty of coal in our joint work sufficient to serve the country to their satisfaction, will be a just cause of cuspition if not of disputes'.[50] Smith was apparently not deterred, for 39 boiler-plates were delivered to him in September 1764. Unspecified fire-engine castings worth £224 followed in November, and then in December he received a cylinder 60 in. x 10 ft weighing 85 cwt., a bottom, a piston with wrought-iron stem and braces, four pipes 16¾-in. x 9 ft, a clack and a working barrel, a sinking-pipe, two buckets and two clacks, a brass sinking-clack and five pipes cost £68.

From that time until the ending of the account book Smith's firm was supplied with regular small consignments, including over 500 boiler-plates.

There is no indication in the account book of the site of the pits, but Donn's map of 1769 shows two engines at Coalpit Heath in a position which would bring them within the boundaries of Westerleigh parish, and a large-scale map of the manor made in 1772 also shows two engines there, one standing on the common, and one in an enclosed field near it.[51] In that year Sir Jarret Smith bought the 70-in. cylinder, piston and bottom of an engine that had been at the brass works at Warmley.[52]

In spite of this there were clearly only two engines at Coalpit Heath until 1789. On 12 June in that year the lords of Westerleigh manor (Henry, Lord Middleton, Thomas Smith, and Mrs Sophia Colston) entered into an agreement with William Bond of Bristol and Thomas Palmer of Stapleton, engineers, who were 'to erect, carry on, and compleatly finish according to the model and other directions to them given by Mr Robert Barber an engine now erecting upon an estate called Serridge'. They were to finish it before 1 October 1790, and during that time they were also to 'do their utmost to put and keep in repair the two engines now used at Coalpitheath ... so that they may be used with safety'. For this they were to be paid at the rate of £1 for a six-day week each, and at least one was to be at work there during the whole period. A lump sum of £20 for the two was payable on completion.[53]

A series of notes give details of the proposed engine:

	yds	ft	ins
Engine Pitt (now sunk) is from Top at Burge down to Coal	94	1	0
A place is to be made in the head near the bottom for the crooked end of the Wind Bore to be put in deeper	0	2	0
The Levill is from Top of Ground where	95	0	0
Water is to be delivered in	23	0	0
This is the height to which the Water from Colliery will be delivered at except a small quantity for the House Purposes	72	0	0

Two Beams will appear from the back part of the House. One of these is to raise the Water occasionally in dry Weather wanted to supply the Reservoir.

The other Beam must work constantly to raise Water from Reservoir to House Cistern. The Water used for Cillinder runs down a Gout in at one end of Reservoir and passes round to the Pump foot wrought by this beam and consequently is proper for to

prevent corroding and by using it repeatly over and over etc. except what exhausts or evaporates there will require but a small quantity of water to be raised from the levill to the Reservoir and not any in Winter or wet weather

The generall outline of Engine fixed upon are To have two sets of Pumps (or Shides) to raise the Water from the Coal up to the levill divided proportionately

Diameter of working Barrells 16 In

Two crooked Wind Bore Pumps each of them to turn aside at feet into their two distinct by Heads, one at bottom, the other at mid way to take in their Water at

These are crooked for the purpose of taking in their Water from behind the Walls of the pitt, so that the Pitt may be kept dry and without a cistern standing in the pitt which is detrimental, also when this Engine Pitt is hereafter wanted to be sunk 33 yards lower it will be found of great advantage.

An undated note shows that the Scrridge engine made a ten-foot stroke eight times a minute and delivered 86 gallons at each stroke. What was called 'the small engine on the common' had a seven-foot stroke, made eight strokes a minute and delivered 28 gallons per stroke from 55 fathoms through 11-in. shides.

A state of Westerleigh Coal Work in 1792 mentions the Serridge engine and the engine on the common as though they were the only two then working, and goes on, 'If the assistance of more engines be needed, it is intended to affix an additional cylinder to that engine which now stands on that part of the common called Ramhill, which cylinder with several of its necessary appendages is now lodged in the old round engine house'.[54] Ram Hill lies slightly to the east of the places where the engines were in 1772, so possibly one had been moved in the intervening 20 years. The spare cylinder may have been the one bought from Warmley, but it is not clear what was meant by fitting an additional cylinder.

One colliery at Coalpit Heath was worked until 1949, and the writer was one of a small group of students who were shown, early in 1950, a remarkable group of steam-engines which had served it for many years. The pumping work was done by a beam engine of Cornish type made by T. and E. Bush of Bristol in 1850–52. It had an 85-in. cylinder with a 10-ft stroke, and the iron beam weighed 22 tons. Another beam-engine made by Bristol Iron Works, with cylinder 30 in. x 6 ft was used for winding at the Mays Hill shaft; other engines included a bench engine which had been at the colliery since 1839, and horizontal fan engines of 1864 and 1891. That was before the days of industrial archaeology, and all these engines went for scrap.

OLDLAND

In 1761 Samuel Creswicke of Hanham granted a coaling lease to Samuel Holbin and Isaac Jefferies of land at Oldland on the side of the hill and adjoining Oldland Brook.[55] Two years later the account book shows that a parcel of fire-engine castings was delivered to Isaac Jefferys and Co.; no place was mentioned, but it seems likely that it was for this work, although Donn's map of 1769 shows no engine on the site. The cost, £337, makes it fairly certain that a whole engine was involved. Later deliveries to this firm were of 9-in. pit barrels in 1764 and 1767, and 139 boiler-plates in 1767.

Another lease of rights in the same district, but west of the brook, was made to Isaac Baugh and the Rev. Charles Lee, both of Bristol, and John Tippet and Peter Bush of Bitton, coalminers, in 1773. It included a covenant to erect a fire-engine with a cylinder not less than 48-in. in diameter. At the determination of the lease the engine and the iron shides were to become the property of the landowner, Charles Bragge.[56]

PUCKLECHURCH PARK COLLIERY

Between 1763 and 1767 seven consignments were made to Samuel Jefferys and Co. of Pucklechurch Park Colliery. Three consisted of boiler-plates (254 in all), and three of pumpwork, while one was un-specified but worth only £58. The total cost of all seven came to £328. The presence of engines at the site is also vouched for by Donn's map of 1769, which shows two engine-houses north of the road from Puckle-church to Lyde Green, near its junction with the road to Parkfield. An undated plan in the archives of the Pleydell-Bouverie family, lords of

34

the manor of Pucklechurch, shows two engines, distinguished as old and new, in the same position, and another plan shows 'the ground from whence the coal was landed out of the old pit by the stable, from 29 to 44 fathoms deep . . . now full of water', in the same area. It was probably made before the engines were erected.

The only contemporary coaling lease in the same archives dates from 1762, but relates to another concern. It was to Thomas Haynes of Wick and Abson, and Samuel Whittuck of Hanham, who were licensed to work coal in Lyde Green Common.[57]

The Parkfield Colliery at Pucklechurch later included the site of the early engines. When it was for sale in 1900 it was pumped by two Cornish engines, one of 60-in. and one of 54-in.[58]

GOLDNEY HOUSE, BRISTOL

Although the account book contains many details of Goldney's personal investment and his dealings with his relatives, there is little reference to his personal activities. One of the few items of personal expenditure comes in January 1759, when he paid George Perry the remaining half of his subscription for 20 sets of his prints of the Dale, the half amounting to £2 10s. This is the well known series of prints of Coalbrookdale by George Perry and T. Vivares.

Goldney spent much time and money in beautifying the grounds of his house. The account book shows that he bought a turned garden 'rowler' from the Dale in 1753. In 1747 he sent a 'Darbyshire slab' in a cast-iron frame to William Lemon, the Cornish mine adventurer, 'in return for what he sent me for the grotto'. What he sent must have been minerals for decorating the walls of the grotto, which still exists. In 1750 another present was sent to Lemon — castings for his own little fire-engine amounting to £5. This was clearly no more than a model, but when Goldney himself acquired his own engine in 1764 it was a fully practical one. Its purpose was to pump water for the fountains in the garden. Goldney set the parts out in great detail and the account is printed in full in the appendix (p. 56). The total cost (at full commercial rates; no reduction for staff or shareholders) was £67 7s. 8d. In August 1765 a few other small pieces and a cast-iron boiler, which weighed about 12 cwt. and cost £9 16s. 3d. were delivered, but the boiler was returned as too heavy. Its replacement weighed 7¼ cwt. All was not yet well, however, for in June 1766, a new cylinder was fitted, the other having been found too small. It was 15 in. x 9 ft 2 in., and with its bottom, piston and stem cost £23. It required a large boiler, of almost 14 cwt., which at last appears to have made all well. The only other recorded delivery for the engine was of two pipes in 1768.

UNKNOWN COLLIERY

The earliest engine recorded in the account book was for Sir Abraham Elton and Co. The parts were delivered in September 1744, and consisted of a cylinder and bottom, a working barrel, two bored barrels, seven unbored barrels and a sinking-pipe in two parts. The combined weight of the cylinder and bottom, 46 cwt., suggest a diameter of about 30 in; the cost in all was £132. The Elton family had widespread industrial interests around Bristol, including the copper works at Conham, and (as the account book shows), powder mills. This engine was no doubt at a colliery. Somewhat earlier, 1733, a firm of the same name was running a coal work called Cowland Hill,[59] This was presumably the same as Cowhorn Hill at Oldland, so the engine may have been there.

UNKNOWN COLLIERY

A firm called Berrow and Edwards received parts of a small engine early in 1747. They consisted of a 24-in. cylinder weighing 24 cwt., a bottom, a piston, a sinking-pipe, and two bored pipes, and cost £63. One partner was probably John Berrow, mayor of Bristol in 1743 and sherrif in 1747, but the site of the engine is not known.

UNKNOWN COLLIERY

In November 1754 an unspecified consignment worth £78 was delivered to Isaac Baugh and Co. It clearly included a cylinder, for early in 1755 a bottom, a piston, a sinking-pipe, a bored pipe, six buckets and clacks, and a damper-plate, worth £30, followed. The cylinder-bottom weighed only 7 cwt., suggesting a diameter of about 30-in. The next consignment to Baugh and Co., in 1760, refers specifically to a colliery; it consisted of a bored working barrel 8-in. in diameter, 20 pit barrels, two buckets and a clack, costing in all £147.

It is not known where this colliery was, but it may have been at Oldland, where Baugh had interests rather later (see p.34). Isaac Baugh was described in Sketchley's directory of 1775 as alderman and merchant, and the same source refers to Baugh, Ames and Co. of 'the gunpowder office, the first floor on the right hand up one pair of stairs, 17 in the Exchange'. The supply of gunpowder-mill castings to Isaac Baugh and Co. is noted in the account book in 1759 and 1764.

UNKNOWN COLLIERY

Two consignments made in 1762 and 1763 to Charles Bragge were specially marked 'for his own work not Nibley Co.' The first was of a bored pipe, two buckets and two clacks, ten pit barrels and a bucket-door pipe, and came to £90. The next included a blast-hole piece and a clack-seat piece, and came only to £11.

Bragge's interests in the coalfield were widespread and it is not possible to be sure where this engine was. It could have been the one at Staple Hill, or the New Level one which Bragge sold in 1779 (see p.37).

KINGSWOOD

In 1735 Richard How Chester was granted a lease of coaling rights in a

36

part of the waste-land of the Forest of Kingswood by his brother, Thomas Chester of Knowle in the parish of Almondsbury. Its favourable terms were because he had spent over £400 in erecting a fire-engine to drain the seams of coal there. This site was described as being in a nook or corner on the south side of merestones which ran from the 'Broad Arrowed Oak' to the liberty of Mangotsfield. No more is known of this engine until 1754, when Richard Chester had evidently given up his lease, and his brother made a new one to Norborne Berkeley of Stoke Gifford, with the use of the fire-engine in his liberty in the Forest of Kingswood in the parish of St Philip and St Jacob.[60]

This lease may have been a renewal, for Berkeley was engaged in coalmining in 1750, when the Bristol newspaper reported the death of Charles Arthur, the superintendent of his coal-work.[61] Another newspaper report in 1748 refers to an assault by a negro on 'the master of the fire-engine and one of the overseers of the cole works in Kingswood';[62] quite probably it was the Chester engine that was meant.

Norborne Berkeley was one of the local coal adventurers who in 1761 joined the Warmley Company, and this led to his financial downfall in 1768. By this time he had proved his right to the barony of Botetort, which had been in abeyance since 1406,[63] and it is by this title that the account book refers to him in 1766 and 1767, when he was supplied with boiler-plates, 216 in all, at a cost of £116.

The position of the Chester engine is not certain, but it could well have been the one marked by Donn in St. George on the south side of the Marshfield road. The Chester family papers contain a lease dated 1819 of 'a cottage called the Fire Engine at the Malmsey Pool adjoining the Bristol to London road in the parish of St George', which probably refers to the same site.[64]

STAPLE HILL

Donn's map shows an engine-house at Staple Hill on the north side of Mangotsfield road. It belonged to a coal work called Staple Hill with veins called Cock, Chick and Hen which belonged to Charles Bragge. In 1779 it was excepted from a lease of general coaling rights in the manor of Mangotsfield, but there is no indication of which firm was working it. The bills of the adjoining Sheppard's work include two for small coal sold to Staple Hill in 1778 and 1779.[65]

Nearby, on the south side of the road, was the coal work called Sheppard's Veins, including Stubbs vein, the Shelly vein, the hard vein and Britton's vein. In 1778 Charles Bragge let the work to a partnership of Zachary Bayly of Lyncombe and Widcombe near Bath, esquire, and Moses Rennolds of Bedminster and Joseph Flook of St George, coal miners. In the following year he sold them what were described as the 'utensills' of a fire-engine which stood at the New Level coal work, which is marked on Taylor's map of 1777 as being nearby. It is not clear what this means. A bill of 1779 refers to part of the utensils, valued by Charles Palmer at £264, but gives no details, and a £10 allowance was made for taking them down. On the whole it seems likely that these included the cylinder and other main parts, which are not mentioned in a further bill running from 1779 to 1781. This was made in detail and appears to give the parts as they were bought piece by piece. None of them were actual engine parts; besides working barrels, shides, buckets and clacks they included a horse-wheel, the engine capstan, the materials of the New Level engine-house, and much general coalpit material such as timber carts, ropes and boring-rods. The total of this second bill was £316. Other papers show that Bayly and Co. styled their concern the Kingswood coal work.[66]

YATE

Donn shows an engine at Yate on the south side of the road from Yate to Nibley, marking the spot as Yate Heath. It has not proved possible to identify this with one of those mentioned in the account book but unplaced.

North of Yate, and mainly in the adjoining parish of Iron Acton, is the area called Engine Common. Taylor's map of 1777 marks it as Antien Heath. Early map-makers were prone to make mistakes — could this be a mistake for Engine Heath? Neither Donn nor Taylor shows an engine-house at the site. A colliery there known as Yate Colliery was worked until 1886. In 1884 it was being pumped by two old engines at shafts called the Old and New Pits. That at the old pit was replaced by a modern one in 1885, but the New Pit engine was still used for two or three hours a day to supply the boilers. When the machinery of the colliery was sold in 1887, the engine was described as a large condensing engine, with a 54-in. cylinder and 7-ft stroke, working two 13-in. lifts of pumps. It could possibly have been of Newcomen type.[67]

CROMHALL

A valuation of the colliery at Cromhall made in 1786 refers to a fire-engine which had cost £700 and could probably be sold for £350. Three years later another valuer considered that it was worth £420, out of a total of £509 for the whole of the colliery's equipment. It would use 720 bushels of coal a week, which worked out at a cost of £312 a year to the adventurers.[68]

COWHORN HILL IN THE PARISH OF BITTON

The manuscripts of the Haynes family of the parish of Wick and Abson contain many papers relating to their involvement in collieries there and in the neighbouring parish of Bitton. In 1758 Thomas Haynes entered into a partnership with Christopher Williams and William Smith to work collieries in Bitton, and two years later obtained an estimate for a very large water-wheel to pump up through 60 fathoms of timber shides. The diameter of the wheel was to be about 44 ft, and the total cost about

£386. Charles Palmer, the engineer, was to receive 2s. 6d a day (of twelve hours) and his men 1s. 8d and 'some time perhaps a mug of drink'. [69]

This water-engine was no doubt meant for a colliery at Cowhorn Hill, which was being worked by a firm called Richard Haynes and Co. in 1792. In that year they decided to fit a second-hand fire-engine and had two inspected. The one they bought came from Hanham, and it seems most likely that the other was there too. The rejected one was the 28-in. engine supplied to Moses Slade and Co. in 1750 (see p.31).

The engine that they bought was probably the other one of the two at Hanham shown on Donn's map of 1769, but it cannot be identified with any of those in the account book. By 1793 it was the property of Harfords and the Bristol Brass and Copper Co., whose bill to Haynes and Co. included the following main items:

	£	s	d
Weigh beam, wrought iron martingels, gudgeon and trow	35	0	0
Cast iron chains with wrought iron bolts 17.1.26 at 6s.	5	4	10
Injection cistern, lead on the inside, cylinder cup and all lead pipes 43.2.24 at 12s.	39	6	10
All the wrought iron belonging to the engine and boylers 38.2.26 at 13s.	25	3	6
1 Piston and stem together 15.0.0 at 9s. 6d.	7	2	6
1 Cylinder 68.1.10 at 14s.	47	16	9
1 Cylinder bottom 18.3.18 at 9s.	8	10	3
1 Hot well 5.2.0 at 9s.	2	9	6
All the steam pipes, sinking pipes and feed pipes 56.3.23 at 8s.	22	15	8
Brass cocks and regulators belonging to the engine and some brasses in the counting house 5.3.25 at 56s.	16	14	6
1 Working barrel for the jack pump 10.0.0 at 14s.	7	0	0
1 Clack piece and blast hole together 6.0.0 at 9s.	2	14	0
1 Capston, wrought iron, brass timber and tile over the same	8	0	0
2 wrought iron boilers with copper tops	70	0	0
Old cast iron that was in the ballance box 25.0.9 at 1s. 6d.	1	17	7

Other small items, including the materials of the engine-house brought the total cost to £339 5s. 6d. In July and August 1793 Thomas Palmer's bill for taking down the engine at Hanham came to £29 15s. 10d. From then on his running bill for himself and his men's work and for small iron-work and tools until the engine was set going on 11 February 1795 came to £165 19s. 11½d A plumber's bill of £24 3s. 6d from Edward Stephens and Co. of Bristol, one of £58 19s. 10d to Leonard and Riddel for cast-iron pitwork, and one of £9 2s. 6d to Richard and

Thomas Haynes and Co. for boiler-plates, brought the recorded expenditure up to well over £600. It is worth noting that the master engineer's daily rate had risen from the 2s. 6d a day of 1760 to 3s. 4d in 1793.

The engine evidently worked successfully, although an accident was noted by the engineer on 14 January 1795 when the pump-rods slipped while the clack was being changed, which kept the engine at a standstill until 18 January at four o'clock in the morning. Shortly afterwards the company bought a complete set of pitwork from a colliery at Cromhall Common consisting of a blast-hole piece, a clack-door piece, a working piece, a bucket door-piece, and 12 plain shides, amounting in all to 138 cwt.[70]

WARMLEY

Braine's *History of Kingswood Forest,* published in 1891, tells us that 'in passing out of the village [Warmley] towards Siston's Common there formerly stood one of those curious old engines made for pumping water with the "open-top" cylinder — slowly, at about two strokes a minute, it ground away night and day for nearly a hundred years'. [71] The position described could correspond with that of the pumping shaft of the Crown Colliery. A sale particular of 1880 shows that the engine on this shaft had a 34-in. cylinder a 6-ft stroke, and worked 80 fathoms of 12-in. pumps arranged in two lifts, said to have been at work 'for years'.[72] In 1891 this colliery was owned by Goldney and Son, the senior partner being Sir Gabriel Goldney of Chippenham. Isaac Taylor's map of Gloucestershire of 1777 shows an engine-house in a position apparently slightly farther from the village, but the scale makes it difficult to be sure.

SOMERSET

The recent history of the Somerset coalfield, by C.G. Down and A.J. Warrington, repeats the statements of Greenwell and McMurtrie in 1864 that the first engine used was perhaps at Paulton Engine Pit, c. 1750.[73] The writers found no firm evidence of the early use of steam-power, so that here the account book is especially valuable in that it refers to at least three engines.

PAULTON

A pit called Brewer's Pit at Paulton was in existence c. 1700, when it was about 100 ft deep and using an eight-ft wheel for drainage.[74] It may have been the same pit which in 1736 was being worked by a partnership of Elizabeth Brewer of Winchester, Robert Houlton of Trowbridge, and Richard Hewish, clothier and Joseph Hill and Jacob Carter, coal miners, all of Paulton. They took a lease of water-rights in Littleton Brook,[75] but water-power probably proved insufficient, for two years later a 21½-in. cylinder was cast at Coalbrookdale for Hewish and Co. of 'Polton'. Dr Mott assigned this tentatively to Polton, Midlothian, but there can be no doubt that the reference was to Paulton.[76]

Parts of another engine for Richard Hewish and Co. were sent to Goldney in 1745. The cylinder was 30 in. in diameter and weighed 33 cwt., and there was a bottom, piston, sinking-pipe in two parts, and two bored pipes. The total cost was £85. Later, between 1758 and 1764, several small and unspecified consignments were sent to Joseph Hill and Co. of Paulton; the total cost was only £66. The last of these was sent via John Shore, engineer.

Donn's map of 1768 shows two engine-houses at Paulton, which seems sufficient confirmation that the engines of 1736 and 1745 were those that gave Paulton Engine Pit its name.

WELTON COLLIERY

Down and Warrington refer to a coaling lease at Welton in 1756, and a further one was made in 1762, James Lansdown of Radstock being one of the grantors and the well-known names of Mogg and Bush among the takers.[77] When a parcel of fire-engine castings for Welton Colliery passed through Goldney's hands in 1762, it was ordered by John Shore on behalf of James Lansdown and Co. The cost was £187.

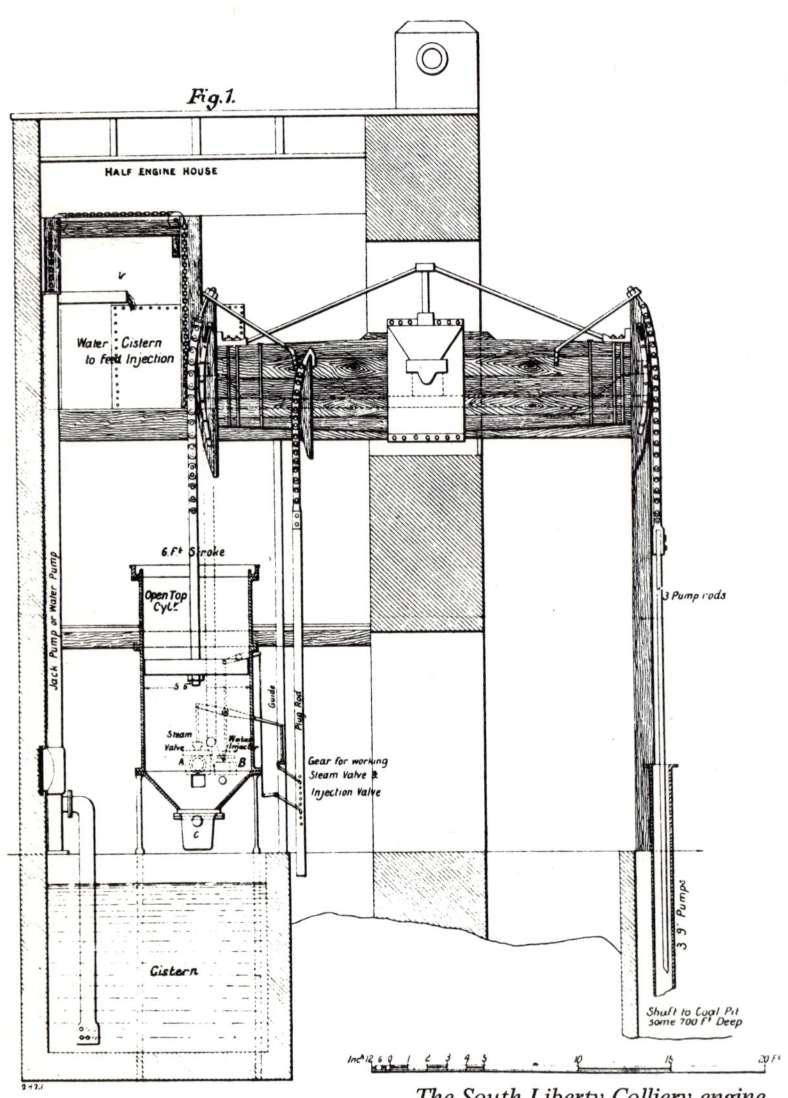

The South Liberty Colliery engine

CLUTTON COLLIERY

Down and Warrington refer to no early colliery at Clutton, but in 1763
an unspecified consignment worth £229 was sent to Joseph Broadribb
and Co. of Hallatrow for Clutton Colliery. The amount makes it virtually
certain that it consisted of parts of an engine, although Donn's map

42

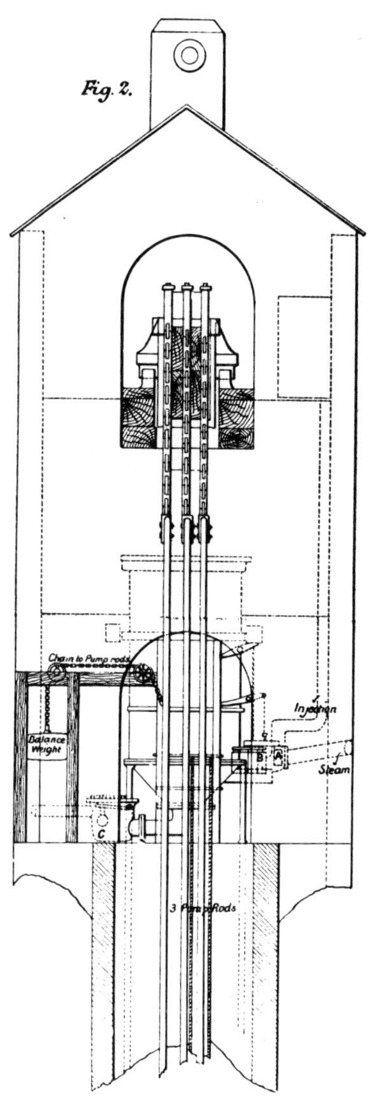

shows no engine-house near Clutton.

Down and Warrington quote references to the purchase of archhead links from the records of New Tyning Pit, Paulton, in 1842, and Old Grove Pit, in 1853 and 1861.[78] This seems conclusive evidence that both were using early engines, either Newcomen or Watt single-acting type, and since they both dated from the eighteenth century, begun in 1791 and 1766 respectively, they probably still had their original engines.[79]

It was on a mine at Radstock that Jonathan Hornblower erected the first compound engine in 1782. It has been said that a second was built at Timsbury in 1784,[80] but Boulton and Watt's statement in opposition to Hornblower's patent in 1792 says categorically that only two had been built, and the second of these must have been the well-documented one at Tincroft Mine, Cornwall, built in 1791.[81]

NAILSEA HEATH COLLIERY

The Smith archives contain an estimate of the numbers and wages of men needed to work this colliery, which by its handwriting appears to date from the late eighteenth century. It includes 'engine men 12s. a week, two men to an engine in winter'.[82]

SOUTH LIBERTY COLLIERY

The last Newcomen engine to work in the west of England and one of the last anywhere, was at this colliery. We are lucky to have photographs and a description of it printed by Bryan Donkin M.I.C.E. in the journal *Engineering* in 1895:[83]

'The engravings on our two-page plate illustrate an interesting old engine which the writer had lately the opportunity of examining on the spot, while working. It was at first thought impossible to take a good photograph of it, as the engine-house is very dark, and there is a floor half-way up, nearly level with the beam, but the managers having removed the roofing and boarding, some fairly good photographs were taken, a few of which we reproduce. The following particulars may be interesting:

The engine was originally supplied with steam from two haystack boilers which about 30 years ago were condemned as unsafe and taken away. Steam has since been obtained from other boilers near, and reduced in pressure for this engine to 2½ lb. The engine is still worked about five hours a day, and six days a week, to keep the South Liberty coal-pit dry. It is probably one of the oldest, if not the oldest, now running. The coal-mine to which the pit belongs is about three miles from the centre of Bristol, and owned by the Ashton Vale Iron Company, Bedminster. The pit, 750 ft. deep, was sunk about 150 years ago, and the engine is said to be of the same age. Assuming this date to be correct, it would bring the construction of the engine to about the year 1745. The mine is still worked, and supplies a certain quantity of coal.

Figures 1 and 2 are engravings of the engine. The cylinder is 5ft. 6in. in diameter and 6 ft. stroke. The piston is packed with old rope. The cylinder is of iron, cast in one piece, with conical-shaped bottom to drain the water, and weighs about 6 tons. The engine, as shown in the

44

The South Liberty Colliery engine in 1895

drawing, has a wooden beam 24 ft. long and about 4 ft. deep, built up of many oak beams trussed together, and works with a curious creaking noise; it weighs, with the iron-work, trusses, gudgeons, etc., about 5 tons. There are three valves, all at the bottom of the cylinder; a circular lift steam valve, a slide valve for injecting water to condense the steam, and a flap valve for the escape of the condensed water. The steam valve is marked A in the drawing, at B is the water injection valve into cylinder, and both are actuated by levers and rods from the beam. At one end of the beam there are three pump-rods raised by chains, as shown, and these lift three water pumps, 9¾ in. in diameter, which deliver one into the other, the first pump being at the bottom of the pit. The three piston-rods are also attached to the beam by three chains, and all the chains work on wooden arcs of circles, as seen in the drawing. There is a jack water pump on the left-hand side of the engine, which lifts water into the cistern, level with the beam, to supply the injection to the bottom of the cylinder. The engine makes about 10 to 11 motor strokes per minute. It works with several inches of water at the top of the piston, so that, if the latter leaks, water only passes through, and thus the vacuum is maintained. No indicator diagrams had ever been taken from the engine, but one has lately been obtained especially to illustrate these notes, and is reproduced. For this purpose the cylinder had to be drilled and gear fitted. The diagram was taken under great difficulties, amid a cloud of steam and coal-dust. The power of this diagram, taking 10 strokes a

45

The South Liberty Colliery engine in 1895

minute, works out at about 52¾ indicated horse-power, a very small power, considering the weight and size of the engine, and compared with present practice. The writer has endeavoured, without success, to find out the cost of the engine.

The old man who attends to the engine is shown in Fig.7; he has driven it since he was a little boy, and had to stand on a block of wood to reach the valve handles. His father and grandfather worked it before him.

In course of time it is hoped that this venerable relic may find a last resting-place in the South Kensington Museum among so many other remarkable antiquities.'

The engine worked for a few years after Donkin wrote, but was not preserved.

The beginning of this colliery may well be dated by a coaling lease of lands in Bedminster from Mrs Ann Smyth of Ashton Court to Jarrett

Smith and others in 1754,[84] and by an advertisement in a Bristol paper in 1755, which announced the making of a new road from Bedminster Bridewell to the new coal mine there 'where coal is sold on as reasonable terms as at any other colliery'.[85] It is possible that it did not have an engine at first, and Donn's map shows no engine-house. The 66-in. cylinder does not correspond with any mentioned in the account book as having been supplied to the Bristol area, and was larger than any sent to any local colliery. It may well represent the colliery practice of the later part of the century.

OTHER ENGINES

SWANSEA

In 1746 Robert Morris and partners built a copper works and 40 cottages on a site adjoining the River Tawe, and two years later established another copper works upstream at Fforest Ucha. They also opened up collieries in the neighbourhood of Treboeth on the west bank of the river.[86] The lease of the first site was renewed in June 1754, and in November of the same year a cylinder was consigned to Morris at Swansea. Its dimensions are not given, but its weight of 30¾ cwt. suggests a diameter of about 30 in. The cylinder bottom, piston, a sinking-pipe and 14 other pipes followed early in 1755. The total recorded cost was £204. The only other recorded consignments were in 1764, unspecified and amounting to only £48.

LLANSAMLET NEAR SWANSEA

Chauncey Townsend, a merchant of Austin Friars in the City of London, was active in exploiting coal-pits in the manor of Kilvay in this parish, and in 1757 set up copper-smelting and refining works on the east bank of the River Tawe.[87] The first recorded consignment to him was an unspecified one of £359, large enough to represent at least a complete engine, in 1763. In 1764 a cylinder 10 ft x 40 in. with piston and bottom, working barrels, pipes etc. were sent, the cost being £194. A larger engine followed in 1766, the cylinder 10 ft x 65 in. and the total cost £432. In 1767 and 1768 six separate consignments of boiler-plates were made, besides pit barrels, a steam-vessel, and other small items, which cost in all £520.

COURT HERBERT, NEATH

In 1766 Thomas Williams of Court Herbert took delivery of a cylinder 54 in. x 10 ft, weighing 70 cwt., and other parts not described, worth £289 in all.

GNOLL NEAR NEATH

An unspecified 'parcel' was delivered to Herbert Mackworth of Gnoll in 1768. A steam-engine is known to have been in use at the colliery there by 1753.[88]

Among the best known of the Newcomen engines was one at Walker Colliery on Tyneside. The first shaft there found coal early in 1762, and two engines were erected, followed by another with a cylinder 74 in x 10 ft 6 in. described as 'the most complete and noble piece of ironwork' which had so far been produced at Coalbrookdale.[89] The weight was 6½ tons. The account book refers in a marginal note to the sending of this cylinder and its bottom (weighing 49 cwt.) in December 1762, but they do not appear to have passed through Bristol. It was this engine which was described by Gabriel Jars, a French metallurgist, in 1765. It had been built by a colliery engineer, William Brown, and was the largest in the area. Jars mentioned that it was worked by three boilers, with another in reserve, all of wrought iron with lead tops except that under the cylinder, which had a copper top. The piston was packed with hemp rope, and it was necessary to use three jets for condensing owing to the great size of the cylinder.[90]

In the same year, 1765, parts of four engines were shipped from Bristol to Isaac Thompson, a close friend of William Brown and a relative of Abraham Darby, who acted as the Coalbrookdàle representative on Tyneside. The cylinders were respectively 64 in. x 10 ft (95 cwt), 75 in. x 10½ ft (106 cwt.), 66 in. by 10 ft (85 cwt.) and 60 in x 10 ft (74 cwt.). Also sent were the necessary bottoms and pistons, four working barrels (one 16 in. diameter, the rest unspecified), and 41 unspecified pipes. The first consignment in January weighed almost 35 tons, in spite of the fact that the 60-in. cylinder had to be left behind because there was no more room in the ship, the *Experiment*. The *Defiance* followed in the autumn with another 20 tons. Goldney noted, however, that the 75-in. cylinder weighed 24 cwt. less than the 74-in. one of 1763.

By comparing the cylinder-sizes with a list of engines in the northern coalfield supplied to Smeaton in 1769, and with a chronological list of engines built there compiled by Dr Raistrick, it is clear that the 64-in. cylinder was for Lambton, Co. Durham, the 75-in. one for Benwell, and the 60-in. one for Long Benton. The Smeaton list mentions no engine with a cylinder of 66-in. so its destination remains unknown.[91]

Dr Raistrick printed extracts from the notebook kept by Christopher Bedlington, William Brown's assistant, giving details of the erection of Benwell engine. The cylinder, which left Bristol some time in September 1765, was on the quay in the north by 19 December. In February he noted that the springs and spring-frame had been made for a 9-ft stroke, but the cylinder would not allow more than 7 ft 6 in., although the account book shows that it was 10 ft 6 in. long. The cylinder was not fixed in position until 13 June 1766. In July they were getting in the pitwork. The working barrel was 22½-in. in diameter, and the pit barrels were of hooped wood.

The engine was started on 28 July, but made a poor beginning. On 1 August Bedlington noted that she would not do more than eight strokes a minute owing to air in the cylinder, and on 8 August he wrote 'Attended the engine all day and at past 11 o'clock drew the water down, but did not keep her down for long, for the smallest abatement of the fires set her up again. Was busy fixing the working gear so that the engine may be herself, when she chooses she still goes no more than 8 strokes a minute . . .' Many other details of his work are given, and it was not until 27 November, when the engine was doing ten strokes of eight ft a minute, that she was satisfactory.[92]

BOW, LONDON

Parts of a small engine, the cylinder 24 in. in diameter, were consigned to Thomas Byrd in 1745. They included also an 'engine tree', a bottom, three bored barrels and one unbored, a sinking-pipe and a piston, and came in all to £96. It has not proved possible to identify Byrd. There were dye-works and distilleries in Bow at this time,[93] but no early reference to the use of an engine in them has been found. Farey speaks of early engines in London only in connection with waterworks, and it is likely that this one was for that purpose.

SCOTLAND

In 1751 a 38-in. cylinder, a piston, two working barrels (9 ft x 9 in. and 9 ft x 9¼ in.), a brass regulator, a jackhead pipe and some other small parts were consigned to William Hare, whose rather imprecise address was simply Scotland. Total cost was £142.

ENGINE PARTS

The principal casting was of course the cylinder, and the following list shows the sizes of the 50 cylinders mentioned in the account book in chronological order. Figures in square brackets represent a guess at the diameter where it was not specified.

	Date	Place or firm	Diameter	Weight cwt.qr.lb	Weight of bottom cwt.qr.lb
1	1744	Near Bristol	[33]	46.1.14 together	
2	1744	Trevenson	[40]	43.1.7	19.0.26
3	1745	Trevenson	[47]	60.0.2	—
4	1745	Paulton	30	33.0.21	9.0.25
5	1745	Bow, London	24	25.1.7	6.0.22
6	1746	Ludgvan Leaze	47	56.0.4	18.1.8
7	1746	Roskear	47	60.2.0	19.1.24
8	1746	Dolcoath	40	46.2.18	17.1.19
9	1746	Lemon and Co.	[40]	48.1.4	16.0.20
10	1747	Near Bristol	[26]	24.2.4	6.1.12
11	1747	Polgooth	[40]	63.2.14 together	
12	1748	Warmley	36	51.3.13	13.0.8
13	1748	Pool Adit	60	90.2.0	41.2.13
14	1748	Drannack	55	71.1.14	31.3.11
15	1749	Warmley	48	67.2.17	17.1.8
16	1749	Lemon and Co.	52	83.0.15	23.2.17
17	1750	Polgooth	54	67.0.21	18.2.18
18	1750	Hanham	28	27.1.21	6.1.0
19	1750	Soundwell	26	24.1.18	5.2.18
20	1750	Chacewater	54	68.0.7	20.0.0
21	1751	Coalpit Heath	42	54.1.7	15.3.14
22	1751	Scotland	38	37.1.21	—
23	1753	Dolcoath	54	62.0.14	25.2.17
24	1753	Nibley	44½	55.3.3	16.3.21

25	1753	Herland	70	116.2.20	48.2.10
26	1753	Wheal Rose	60	99.3.7 together	
27	1754	Near Bristol	[30]	—	7.2.14
28	1754	Swansea	[30]	30.3.0	?
29	1756	North Downs	60	93.0.23	30.3.14
30	1756	Soundwell	33	37.1.20	9.1.20
31	1758	Wheal Virgin	60	89.1.0	30.3.0
32	1761	Warmley	74½	128.0.18	51.2.12
33	1763	Wheal Oula	70	110.2.5	—
34	1763	Poldice	60	81.1.0	30.2.25
35	1764	Llansamlet	40	48.2.16	13.1.11
36	1764	Goldney House	11	6.1.21	1.0.0
37	1764	Coalpit Heath	60	85.1.12	22.2.18
38	1765	Long Benton	60	74.2.6	30.3.13
39	1765	Lambton	64	95.2.12	31.3.11
40	1765	Wheal Virgin	60	80.3.12	—
41	1765	Warmley	70	101.2.14	37.0.25
42	1765	Tresavean	60	84.3.24	29.3.23
43	1765	Benwell	75	106.0.6	49.0.0
44	1765	Newcastle	66	85.3.16	36.1.14
45	1766	Llansamlet	65	88.2.15	32.2.17
46	1766	Chacewater	66	96.1.0	37.1.10
47	1766	Goldney House	15	13.1.9	1.2.5
48	1766	Court Herbert	54	70.1.9	—
49	1767	Wheal Virgin	60	82.1.12	26.1.13
50	1768	New Dolcoath	63	83.0.5	30.1.5

Leaving aside the two very small cylinders supplied to Goldney himself, we see that 11 cylinders below 40-in. diameter were supplied between 1744 and 1756; none of these went to Cornwall, and the majority of them were for collieries. Of the next size, 40-49 in., seven went to Cornish mines, all in the early period, 1744—47, while the four supplied elsewhere were between 1749 and 1764. The same picture of Cornish practice outstripping other users comes from the 50—59-in. group, of which five went to the county between 1748 and 1753, while the only other one supplied (to South Wales) came in 1766. Of the 60—69-in. group ten went to Cornwall between 1748 and 1768, while only four went elsewhere between 1764 and 1766. Three of these wer

only four went elsewhere between 1764 and 1766. Three of these were part of a group of four large cylinders sent to Newcastle-upon-Tyne in 1765; the other was of 75-in. Only four other cylinders of 70-in. and above were supplied, two to Cornwall in 1753 and 1763, and two to the Bristol area in 1761 and 1765. Clearly at this time, as in the nineteenth century, the problems of draining the Cornish mines stimulated their proprietors to boldness in fitting engines of the largest size available. The 70-in. engine of 1753 was probably the largest built anywhere at that time. Cylinders and bottoms were charged at the rate of all bored work, 30s per cwt.

Apart from the cylinder and its bottom, there was great variation in the parts which were supplied for individual engines. Two which were almost always mentioned were the piston, which was turned and had a wrought-iron stem or stems, and the sinking-pipe. The latter was apparently Goldney's name for the eduction pipe which carries off the hot water from the cylinder bottom. It was often described as being in two or three parts, which corresponds to early diagrams, and the name sinking-pipe is used on the Barney diagram of 1719. It is not clear why this pipe was usually charged at the bored-work rate.

Many engines were supplied only with the four parts so far mentioned. Several instances occur of a brass piston-ring, which was charged at 1s. or 1s 4d. a lb. and is described in two entries as being turned. Farey's very full descriptions of Newcomen engines mention no such part; according to him the piston had a flange on its upper surface, and between that and the cylinder-walls a packing of oakum mixed with horse-dung was rammed and kept down by cast-iron weights.[94] From the material of the piston-rings and their being turned it is hard to escape the conclusion that they were meant to form a working rim to the piston. Another part not mentioned by Farey, at least by Goldney's name, was the maundrell, which was turned. Only a few instances occur of the supply of the various engine-valves and pipes, and the account book largely bears out the accepted view that local men were able to supply most engine-parts except the major castings. Exceptions were Goldney's own engine and the 1761 Warmley engine, and the accounts for these are transcribed below.

The book also reveals very clearly the great quantities of pitwork supplied to many mines. Few engines were sent without the bored working barrels. In the early years the book reveals several instances of the supply of brass working barrels, which were sometimes thought necessary if the local water was corrosive. The price of these, at 1s. a lb. could be very great; three sent to Drannock in 1748 weighed 49 cwt. and cost £275. A spelter working barrel sent to Cornwall in 1747 was charged at 10d. a lb., but returned as faulty. Iron working barrels occur throughout the book, charged at 30s. a cwt. The absence of later references to brass barrels does not mean that they had gone out of fashion entirely, for Farey quotes a bill for an engine built in 1775 which had two brass barrels.[95] Perhaps they were supplied from another source in the 1760s.

The other parts of the pitwork, both the plain pit barrels, the special pipes for the clacks and bucket-doors with their door-plates, and the blast-hole pieces were frequently supplied at 18s. a cwt., as were the buckets and clacks, which were occasionally of brass. The book records regular consignments of new pitwork to mines, especially in Cornwall, and emphasizes in this field rather more than in the actual engine-parts the great reliance of the mine adventurers on parts from Coalbrookdale.

One field in which Cornwall was, however, generally self-reliant was in the supply of boiler-plates. These did not occur until the last five years of the account book, but in that period were supplied in great quantity to the Bristol area and South Wales. Only one instance occurs of their being supplied to Cornwall, and Mr Harris shows that there was a specialised boiler-plate maker in the county in 1771.[96] Boiler plates, being smith's work, were comparatively expensive at 28s. a cwt., but other parts such as fire-doors and their frames, and damper-plates, were 12s. There are several references to damper-plates, although Farey says that they were unknown in Smeaton's time.[97]

The account book contains some references to the local engineers who built and maintained the engines. The activities of George John of Camborne are illustrated in Mr Harris's article, and several small consignments were sent to him 1753—56. John Dudley of Redruth occurs 1758—61. In the Bristol area Charles Palmer worked in the 1760s, and must have been the predecessor of Thomas Palmer of Stapleton whose activities are illustrated by the later Gloucestershire papers. John Shore, engineer of Paulton, is mentioned in 1764.

EXTRACTS FROM THE ACCOUNT BOOK

THE WARMLEY ENGINE OF 1761

April	Recd. from the Dale per Beard's Trowe as per AD's Letter of March 31st For Warmley Co. 1 Cylinder 74½ In. diamr. 10 ft. long 128.0.18			
Mind Cylinder Bottom	4 Working Barrels, bor'd 9 ft. long, 24 Inches diamr. <u>127.1.26</u>			
	255.2.16 at 30/–	£383	9	3½
	4 Pitt Barrels 9½ ft. long 29¾ in. diamr.			
	133.1.24 at 18/–	120.	2	4¼
	1 Hot Well 8.0.14 at 12/– cast at 8.0.4 in Dale Invoyce	4	17	6
	3 Engine Chains and 3 Hooks 29.1.10 at 42/–	61	12	3
	Screw Pins, Burs, and Rings, 32 of each			
	2.0.13 at 56/–	5	18	6
	The Cylinder Bottom as per AD's letter			
	Ap.21st 51.2.12 at 30/–	77	8	2½
April 22	Recd. from the Dale per Beard's Trowe as per A.Darby's letter of April 21st, for Warmley Co.			
	1 Piston turn'd 32.0.0. at 12/–	19	4	0
	3 Wrott. Stems, 5 Braces, 3 Cotters and Sword			
	8.3.19 at 6d. per lb.	24	19	6
	1 Breeches Pipe and Steam Vessel top 13.0.27			
	1 steam Vessell <u>14.3.7</u>			
	28.0.6 at 16/–	22	8	10¼
	1 Lead Ring 0.1.2 at 2d. per lb		5	0
	1 Brass Regulator 3.0.17 at 16d. per lb	23	10	8
	Iron screw pins 0.3.7 at 6d. per lb	2	5	6
	3 Yards of Flannel		2	3
May 11	For Warmley Co. 4 Clack and Blast hole Pipes and 1 house			
	Water Pipe 150.0.10 at 18/–	135	1	7¼

1 Jack head bor'd 12.1.7
1 Sinking pipe in 2 parts   8.3.4.
                          21.0.11 at 30/–   31 12 11
1 Clack Seat Plate 2.2.10 at 12/–   1 11 0¾

29 May   1 Brass Sinking Clack 64 lb and ½
               1 Steam Clack 30 lb. and ½
               1 Snifting do.   29 lb and ½
                     124½ lb at 1/4   8 6 0

THOMAS GOLDNEY'S ENGINE, 1764–6

1764   Recd. from the Dale per Dale Trowe as per
Novr.5th   R.Reynold's lettr. 9 br   9th
              Fire Engine Castings etc for my Self viz:
              A Cylinder 8 ft. long 11 Inches diamr. 6.1.21
              A Bottom to it   1.0.0
                          7.1.21 at 30/– £11. 3 1½
              1 Piston turn'd 0.1.3½ at 12/–   3 4
              Wrought Stem in ditto 0.0.14½ at 6d.   7 3
              1 Steam Vessel and Top 1.1.27 at 18/– 1. 6 10
              Brass regulator   0.0.18¼ at 1/4   1 4 4
              5 Cast Chains   4.3.3. at 12/–   2 17 4
              Wrought Iron Pins and Cotters 1.0.3 at 6d. 2 17 6
              Screws and Burrs   0.0.24¼ at 6d.   12 1½
                                   £20 11 10

              Recd. from the Dale as per Richd. Reynold's
              letter of Decr. 25th 1764
              For my own Fire Engine as follows.
              3 Working Barrels 7.0.0., 2.3.0 and 1.3.22
              1 Sinking Pipe 1.1.0
               4 Buckets and Clacks 1.0.26
                      14.0.20 at 30/–   £21 5 4
              3 Clack Seat Pipes 9, 3½, and 2½ Inches diamr.
              2 Elbow Pipes 4 by 4, and 5⅓ by 4
              1 Plain Pipe 9 by 3½
               5 Plain Pipes 5¼ by 4, 1 ditto 5⅓ by 6
                      15.3.2 at 18/–   14 3 10
              6 Rods, 2 Swords, 102 Screws and Burs
              2 Clack Swords, and 1 Clasp
                     3.0.2 at 6d.   8 9 0
              1 Damper Plate 0.2.22 at 12/–   8 4
              4 Brass Buckets and Clacks
              1 Injection Cock 37 lb at 1/4   2 9 4
                                 £46 15 10

1765　　　　　Recd. from the Dale per Saml. Jones's Trow, as per
August 5th　　R.Reynolds's Letter of Augst. 6th 1765
　　　　　　　For my own Fire Engine
　　　　　　　1 Working Barrel 9 ft. long, 3½ inches diamr.
　　　　　　　　　　　　　　2.3.0 at 30/−　　　　£4　2　6
　　　　　　　A Brass Bucket and Clack 6lb at 1/4　　　8　0
　　　　　　　1 Bucket Rod and wrot. iron 29½ lb at 6d　　14　9
　　　　　　　　　　　　　　　　　　　　　　　　£5　5　3

29 August　　For my own Fire Engine
1765　　　　a Boiler 11.3.18 at 16/−　　　　£9　10　7
　　　　　　　68 holes drilled at 1d.　　　　　　　5　8
　　　　　　　　　　　　　　　　　　　　　　9　16　3

　　　　　　　　Retd. as too heavy and recd. another
1765　　　　For my self a Cast Iron Boiler in the place
Octobr 14　　of that return'd 7.1.10 at 16/−　　£5　17　5
　　　　　　　69 holes at 1d.　　　　　　　　　　5　9
　　　　　　　　　　　　　　　　　　　　　　6　3　2

1766　　　　For my own Fire Engine a larger
18th June　　Cylinder than the first (which was
　　　　　　　found too small), viz. 9ft 2 in. long
　　　　　　　and 15 Inches diameter, 13.1.9
　　　　　　　A Bottom to it　　　　　　1.2.5
　　　　　　　　　　　　　　14.3.14 at 30/−　　£22　6　3
　　　　　　　A Piston Plate turn'd 0.2.11 at 12/−　　7　2
　　　　　　　A wrot. Iron Stem to it 20lbs at 6d.　　10　0
　　　　　　　　　　　　　　　　　　　　　　£23　3　5

August 30th Sent from the Dale as per R.Reynold's
1766　　　　Letter of Septr. 2nd 1766
　　　　　　　For my own Fire Engine a Boiler
　　　　　　　　　　　　　　13.3.10 at 16/−　　£11　1　5
　　　　　　　91 holes　　　　　　　　　　　　7　7
　　　　　　　　　　　　　　　　　　　　　　£11　9　0

NOTES

No specific references to the Goldney account book are given, since it is in chronological order and the entries mentioned could easily be found by date. In quoting sums of money and weights from it only the £ and cwt, figures are generally given. The following abbreviations are used:

TNS   *Transactions of the Newcomen Society*
BRO   Bristol Record Office
GRO   Gloucestershire Records Office
Farey   John Farey, *A Treatise on the Steam Engine*, 1827, vol.i (reprinted, 1971)

1. P.K.Stembridge, *Goldney, a House and a Family*, 1969; A.Raistrick, *Dynasty of Ironfounders*
2. B.Trinder, *The Industrial Revolution in Shropshire*, pl.4
3. J.S.Allen, 'The Introduction of the Newcomen Engine, 1710-1733', *TNS* xlii and xliii
4. Farey, 230
5. Allen, *TNS* xlii, 170
6. R.A.Mott, 'The Newcomen Engine in the Eighteenth Century', *TNS* xxxv
7. Trinder, op. cit., 162
8. Mott, *TNS* xxxv
9. A.K.Hamilton Jenkin, *The Cornish Miner*, 99
10. Allen's list in *TNS* xlii, nos. 25 and P4, also addendum to list in *TNS* xliii, no.25
11. T.R.Harris, 'Engineering in Cornwall before 1775', *TNS* xxv, 115
12. 14 Geo. II c.41
13. John Rowe, *Cornwall in the Age of the Industrial Revolution*, 1953, 42
14. Farey, 237
15. D.B.Barton, *A History of Copper Mining in Cornwall and Devon*, 1968, 26–27; Harris, *TNS* xxv, 121
16. Pryce, *Mineralogia Cornubiensis*, 1778, 313
17. D.B.Barton, *The Cornish Beam Engine*, 1969, 23
18. Harris, *TNS* xxv, 115

19. Ibid.
20. Barton, *Cornish Beam Engine*, 270
21. Farey, 190–204
22. Barton, *Copper Mining*, 30
23. Rowe, *op. cit.*, 18
24. Barton, *Copper Mining*, 31–2
25. A.K.Hamilton Jenkin, *Mines and Miners of Cornwall*, iii, 19
26. Barton, *Cornish Beam Engine*, 270
27. Ibid.
28. Jenkin, *Mines and Miners*, vi, 7–8
29. Barton, *Cornish Beam Engine*, 270
30. W.Borlase, *Natural History of Cornwall*, 1758, 173
31. Jenkin, *Cornish Miner*, 200
32. Barton, *Cornish Beam Engine*, 270
33. Harris, *TNS* xxv, 120–122; A.Titley, 'The Account Books of Richard Trevithick Senior', *TNS* xi, 30–33
34. Harris, *TNS* xxv, 117–119; Rowe, *Cornwall in the Age of the Industrial Revolution*, 52
35. Barton, *Cornish Beam Engine*, 270
36. H.L.Douch, *Old Cornish Inns*, 1966, 184
37. Barton, *Cornish Beam Engine*, 270
38. Ibid.
39. Harris, *TNS* xxv, 121–2.
40. D.Vintner, *Some Coalpits in the Neighbourhood of Bristol and Kingswood*, 1964
41. GRO D 1844 C17
42. Joan Day, *Bristol Brass*, 1973, 81
43. Ibid. 84
44. Ibid. 91
45. GRO D421 T39
46. GRO D421 T110
47. A.Emlyn Jones, *History of Mangotsfield and Downend*, 1899, 223
48. BRO HA/B/5
49. BRO AC/AS/97/3, AC/AS/92
50. BRO AC/AS/93
51. BRO AC/PL/85/1
52. BRO AC/AS/97/3
53. BRO AC/AS/97/9
54. BRO AC/AS/97/12
55. BRO 206/17e
56. GRO D421/T38
57. Wilts. RO 490/555
58. BRO 9492/49
59. GRO D421 E62
60. GRO D674a F1, T126

61. *Bristol Journal* 24 Feb. 1750
62. J.Latimer, *Annals of Bristol in the Eighteenth Century*, 1893, 273
63. B.Little, 'Norborne Berkeley: Gloucestershire Magnate'
     Virginia Magazine of History and Biography, lxiii; G.E.C.
     Complete Peerage
64. GRO D674a T220
65. GRO D421 T110, E21
66. GRO D421 T109, E21
67. Wilts RO 515/39, 48
68. GRO D340 T49, D421 E4A
69. BRO HA/B/18, HA/B/5
70. BRO HA/B/5, HA/B/3
71. A.Braine, *History of Kingswood Forest*, 1891
72. Wilts. RO 515/81
73. C.G.Down and A.J.Warrington, *The History of the Somerset Coal-field*, n.d. *c.*1973, 89
74. Somerset RO DD/MGG Box 2
75. Ibid.
76. Raistrick, *Dynasty of Ironfounders*, 295; Mott, *TNS* xxxv
77. Somerset RO DD/MGG Box 2
78. Op. cit., 92, 94
79. Somerset RO DD/RM Box 17
80. Jenkins, 'Jonathan Hornblower and the Compound Engine',
     TNS xi
81. Farey, 387, 389
82. BRO AC/AS/97/12
83. *Engineering,* 25 October 1895
84. BRO 4168/23
85. J.Latimer, *Annals of Bristol in the Eighteenth Century*, 313
86. William Rees, *Industry before the Industrial Revolution*, ii, 574
87. Ibid. 571, 576
88. Ibid. 500
89. Raistrick, *Dynasty of Ironfounders*, 145
90. Singer, Holmyard, Hall and Williams, *History of Technology*, iv,
     178–9
91. A.Raistrick, 'The Steam Engine on Tyneside', *TNS* xvii, 155
92. R.L.Galloway, '*Annals of Coal Mining*, 1891, i, 262; Raistrick,
     TNS xvii, 149–152
93. Inf. from London Borough of Tower Hamlets.
94. Farey, 203
95. Ibid, 232
96. *TNS* xxv, 120
97. Farey, 146

INDEX